Basic Math
and
Pre-Algebra

by
Jerry Bobrow, Ph.D.

Cliffs Notes
INCORPORATED
LINCOLN, NEBRASKA   68501

*T*he Cliffs Notes logo, the names "Cliffs," "Cliffs Notes," and "Cliffs Quick Review," and the black and yellow diagonal stripe cover design are all registered trademarks belonging to Cliffs Notes, Inc., and may not be used in whole or in part without written permission.

Cover photograph by Stephen Johnson/Tony Stone Images

FIRST EDITION

© *Copyright 1995 by Jerry Bobrow*

All Rights Reserved
Printed in U.S.A.

ISBN 0-8220-5307-1

CONTENTS

PRELIMINARIES 1
  Groups of Numbers 1
  Ways to Show Multiplication and Division 2
    Ways to show multiplication 2
    Ways to show division 2
  Multiplying and Dividing Using Zero. 2
  Common Math Symbols 3
  Properties of Basic Mathematical Operations. 3
    Some properties (axioms) of addition 3
    Some properties (axioms) of multiplication 5
    A property of two operations 6
  Grouping Symbols and Order of Operations. 7
    Parentheses ( ), brackets [ ], and braces { } 7
    Order of operations 8

WHOLE NUMBERS 11
  Place Value. 11
  Expanded Notation 12
  Rounding Off 12
  Estimating Sums, Differences, Products, and Quotients ... 13
    Estimating sums 13
    Estimating differences. 14
    Estimating products 14
    Estimating quotients. 15
  Divisibility Rules 15
  Factors, Primes, Composites, and Factor Trees 17
    Factors 17
    Prime numbers 18
    Composite numbers 19
    Factor trees 19

DECIMALS **21**
Definition of the Decimal System 21
Using the Place Value Grid. 22
    Expanded notation 22
    Writing decimals. 22
    Comparing decimals 23
Rounding Decimals 24
Decimal Computation 25
    Adding and subtracting decimals 25
    Multiplying decimals. 26
    Dividing decimals 27
Estimating Sums, Differences, Products, and Quotients . . . 28
    Estimating sums 28
    Estimating differences. 29
    Estimating products 29
    Estimating quotients 30
Repeating Decimals. 31

FRACTIONS **33**
Proper and Improper Fractions. 33
Mixed Numbers. 34
    Changing improper fractions to mixed numbers 34
    Changing mixed numbers to improper fractions 34
Renaming Fractions. 35
    Equivalent fractions 35
    Reducing fractions 35
    Enlarging denominators 36
Factors 37
    Common factors 38
    Greatest common factor 38
Multiples 39
    Common multiples 39
    Least common multiple 39

CONTENTS

Adding and Subtracting Fractions 40
  Adding fractions 40
  Subtracting fractions 42
Adding and Subtracting Mixed Numbers 42
  Adding mixed numbers 42
  Subtracting mixed numbers 43
Multiplying Fractions and Mixed Numbers 45
  Multiplying fractions 45
  Multiplying mixed numbers 46
Dividing Fractions and Mixed Numbers 46
  Dividing fractions 46
  Dividing complex fractions 47
  Dividing mixed numbers 48
Simplifying Fractions and Complex Fractions 49
Changing Fractions to Decimals 51
Changing Terminating Decimals to Fractions 53
Changing Infinite Repeating Decimals to Fractions 53

PERCENT **57**
Changing Percents, Decimals, and Fractions 57
  Changing decimals to percents 57
  Changing percents to decimals 57
  Changing fractions to percents 58
  Changing percents to fractions 58
  Important equivalents that can save you time 59
Applications of Percents 60
  Finding percent of a number 60
  Find what percent one number is of another 61
  Finding a number when a percent of it is known 63
  Percent—proportion method 64
  Finding percent increase or percent decrease 67

INTEGERS AND RATIONALS
(SIGNED NUMBERS) **69**

Integers 69
   Number lines 69
   Addition of integers 69
   Subtraction of integers 71
   Minus preceding parenthesis 72
   Multiplying and dividing integers 74
   Absolute value 75
Rationals (Signed Numbers Including Fractions) 76
   Negative fractions........................... 76
   Adding positive and negative fractions 76
   Adding positive and negative mixed numbers....... 77
   Subtracting positive and negative fractions 77
   Subtracting positive and negative mixed numbers..... 78
   Multiplying positive and negative fractions 79
   Canceling................................... 79
   Multiplying positive and negative mixed numbers..... 80
   Dividing positive and negative fractions........... 81
   Dividing positive and negative mixed numbers 82

POWERS, EXPONENTS, AND ROOTS. **83**

Powers and Exponents........................... 83
   Exponents 83
   Negative exponents.......................... 83
   Squares and cubes........................... 84
   Operations with powers and exponents 85
Square Roots and Cube Roots. 87
   Square roots................................ 87
   Cube roots 88
   Approximating square roots................... 89
   Simplifying square roots 90

CONTENTS

POWERS OF TEN AND SCIENTIFIC NOTATION 93
Powers of Ten 93
    Multiplying powers of ten. 94
    Dividing powers of ten. 94
Scientific Notation 95
    Multiplication in scientific notation 96
    Division in scientific notation 97

MEASUREMENT 99
Measurement Systems 99
    English system 99
    Metric system 100
Converting Units of Measure. 101
Precision. 103
Significant Digits 105
Calculating Measurements of Basic Figures 107
    Perimeter of some polygons—squares, rectangles,
        parallelograms, trapezoids, and triangles 107
    Area of some polygons—squares, rectangles,
        parallelograms, trapezoids, and triangles 108
    Circumference and area of a circle 114

GRAPHS. 117
Bar Graphs 117
Line Graphs 122
Circle Graphs, or Pie Charts 124
Coordinate Graphs. 126

CONTENTS

PROBABILITY AND STATISTICS **131**
    Probability. 131
        Combinations 134
        Permutations 135
    Statistics 137
        Measures of central tendencies. 137

NUMBER SERIES **141**
    Arithmetic progressions 141
    Geometric progressions 142

**VARIABLES, ALGEBRAIC EXPRESSIONS, AND
SIMPLE EQUATIONS** **143**
    Variables and Algebraic Expressions 143
        Variables 143
        Algebraic expressions 143
        Evaluating expressions. 144
    Solving Simple Equations 145
        Addition and subtraction equations. 146
        Multiplication and division equations 147
        Combinations of operations 149

WORD PROBLEMS **159**
    Solving Process 159
    Key Words 160

Groups of Numbers

In doing basic math, you will work with many different groups of numbers. The more you know about these groups, the easier they are to understand and work with.

- **Natural or counting numbers.** The numbers 1, 2, 3, 4, . . . are called **natural** or **counting numbers.**

- **Whole numbers.** The numbers 0, 1, 2, 3, . . . are called **whole numbers.**

- **Odd numbers. Odd numbers** are whole numbers not divisible by 2: 1, 3, 5, 7, . . .

- **Even numbers. Even numbers** are whole numbers divisible by 2: 0, 2, 4, 6, . . .

- **Integers.** The numbers . . . −2, −1, 0, 1, 2, . . . are called **integers.**

- **Negative integers.** The numbers . . . −3, −2, −1 are called **negative integers.**

- **Positive integers.** The natural numbers are sometimes called the **positive integers.**

- **Rational numbers.** Fractions, such $\frac{3}{5}$ or $\frac{7}{8}$, are called **rational numbers.** Since a number such as 5 may be written as $\frac{5}{1}$, all *integers* are *rational numbers.* All rational numbers can be written as fractions a/b, with a being an integer and b being a natural number. Both terminating (such as .5) and repeating (such as .333 . . .) decimals are also rational numbers, since they can be written as fractions in this form.

- **Irrational numbers.** Another type of number is an **irrational number.** Irrational numbers *cannot* be written as fractions

a/b, with a being an integer and b being a natural number. $\sqrt{3}$ and π (the Greek letter pi) are examples of irrational numbers.

Ways to Show Multiplication and Division

Ways to show multiplication. There are several ways to show multiplication. They are

multiplication sign: $4 \times 3 = 12$

multiplication dot: $4 \cdot 3 = 12$

two sets of parentheses: $(4)(3) = 12$

one set of parentheses: $4(3) = 12$ or $(4)3 = 12$

a number next to a variable (letter): $3a$ means 3 times a

two variables (letters) next to each other: ab means a times b

Ways to show division. There are several ways to show division. They are

division sign: $10 \div 2 = 5$

fraction bar: $\frac{10}{2} = 5$ or $10/2 = 5$

Multiplying and Dividing Using Zero

Zero times any number equals zero.

$$0 \times 2 = 0$$

$$8 \times 2 \times 3 \times 6 \times 0 = 0$$

Likewise, zero divided by any number is zero.

$$0 \div 3 = 0$$

$$\frac{0}{7} = 0$$

Important note: *Dividing by zero is "undefined" and is not permitted.* For example, $\frac{4}{0}$ is not permitted because there is no such answer. The answer *is not zero.*

Common Math Symbols

$=$ is equal to

$\neq$ is not equal to

$>$ is greater than

$<$ is less than

$\geq$ is greater than or equal to (also written $\geqq$)

$\leq$ is less than or equal to (also written $\leqq$)

$\not>$ is not greater than

$\not<$ is not less than

$\ngeq$ is not greater than or equal to

$\nleq$ is not less than or equal to

$\approx$ is approximately equal to (also written $\doteq$)

Properties of Basic Mathematical Operations

Some properties (axioms) of addition.

- **Closure** is when all answers fall into the original set. If you add two even numbers, the answer is still an even number ($2 + 4 = 6$); therefore, the set of even numbers is *closed* under addition (has closure). If you add two odd numbers, the answer is not an odd number $(3 + 5 = 8)$; therefore, the set of odd numbers is *not closed* under addition (no closure).

- **Commutative** means that the *order* does not make any difference.

$$2 + 1 = 1 + 2$$

$$a + b = b + a$$

Note: Commutative does *not* hold for subtraction.

$$3 - 2 \neq 2 - 3$$

$$a - b \neq b - a$$

- **Associative** means that the *grouping* does not make any difference.

$$(2 + 3) + 4 = 2 + (3 + 4)$$

$$(a + b) + c = a + (b + c)$$

The grouping has changed (parentheses moved), but the sides are still equal.

Note: Associative does *not* hold for subtraction.

$$4 - (2 - 1) \neq (4 - 2) - 1$$

$$a - (b - c) \neq (a - b) - c$$

- The **identity element** for addition is 0. Any number added to 0 gives the original number.

$$5 + 0 = 5$$

$$a + 0 = a$$

- The **additive inverse** is the opposite (negative) of the number. Any number plus its additive inverse equals 0 (the identity).

  $3 + (-3) = 0$; therefore, 3 and -3 are additive inverses

  $-4 + 4 = 0$; therefore, -4 and 4 are additive inverses

  $a + (-a) = 0$; therefore, a and $-a$ are additive inverses

Some properties (axioms) of multiplication.

- **Closure** is when all answers fall into the original set. If you multiply two even numbers, the answer is still an even number ($2 \times 4 = 8$); therefore, the set of even numbers *is closed* under multiplication (has closure). If you multiply two odd numbers, the answer is an odd number ($3 \times 5 = 15$); therefore, the set of odd numbers *is closed* under multiplication (has closure).

- **Commutative** means that the *order* does not make any difference.

$$4 \times 3 = 3 \times 4$$

$$a \times b = b \times a$$

Note: Commutative does *not* hold for division.

$$12 \div 4 \neq 4 \div 12$$

- **Associative** means that the *grouping* does not make any difference.

$$(2 \times 3) \times 4 = 2 \times (3 \times 4)$$

$$(a \times b) \times c = a \times (b \times c)$$

The grouping has changed (parentheses moved), but the sides are still equal.

Note: Associative does *not* hold for division.

$$(8 \div 4) \div 2 \neq 8 \div (4 \div 2)$$

- The **identity element** for multiplication is 1. Any number multiplied by 1 gives the original number.

$$5 \times 1 = 5$$
$$a \times 1 = a$$

- The **multiplicative inverse** is the **reciprocal** of the number. Any number multiplied by its reciprocal equals 1.

$2 \times \frac{1}{2} = 1$; therefore, 2 and $\frac{1}{2}$ are multiplicative inverses,

or reciprocals

$a \times \frac{1}{a} = 1$; therefore, a and $\frac{1}{a}$ are multiplicative inverses,

or reciprocals

(provided $a \neq 0$)

A property of two operations. The **distributive property** is the process of distributing the number on the outside of the parentheses to each term on the inside.

$$2(3 + 4) = 2(3) + 2(4)$$
$$a(b + c) = a(b) + a(c)$$

Note: You cannot use the distributive property with only one operation.

$$3(4 \times 5 \times 6) \neq 3(4) \times 3(5) \times 3(6)$$
$$a(bcd) \neq a(b) \times a(c) \times a(d) \quad \text{or} \quad (ab)(ac)(ad)$$

Grouping Symbols and Order of Operations

Parentheses ( ), brackets [ ], and braces { }. Parentheses, brackets, and **braces** are used to group numbers or variables (letters).

The most commonly used grouping symbols are **parentheses.** All operations inside parentheses must be done before any other operations.

Example 1: Simplify $4(3 + 5)$.

$$4(3 + 5) = 4(8)$$
$$= 32$$

Example 2: Simplify $(2 + 5)(3 + 4)$.

$$(2 + 5)(3 + 4) = (7)(7)$$
$$= 49$$

Brackets and **braces** are less commonly used grouping symbols and should be used after parentheses. Parentheses are to be used first, then brackets, then braces: $\{ [( )] \}$. Sometimes, instead of brackets or braces, you will see the use of larger parentheses.

Example 3: Simplify $((2 + 3) \cdot 4) + 1$.

$$((2 + 3) \cdot 4) + 1 = ((5) \cdot 4) + 1$$
$$= (20) + 1$$
$$= 21$$

An expression using all three grouping symbols would look like this.

$$2\{1 + [4(2 + 1) + 3]\}$$

Example 4: Simplify $2\{1 + [4(2 + 1) + 3]\}$.

Notice that you work from the inside out.

$$
\begin{aligned}
2\{1 + [4(2 + 1) + 3]\} &= 2\{1 + [4(3) + 3]\} \\
&= 2\{1 + [12 + 3]\} \\
&= 2\{1 + [15]\} \\
&= 2\{16\} \\
&= 32
\end{aligned}
$$

Order of operations. If multiplication, division, powers, addition, parentheses, etc., are all contained in one problem, the **order of operations** is as follows.

1. parentheses
2. exponents
3. multiplication ⎫
4. division   ⎭ whichever comes first left to right
5. addition   ⎫
6. subtraction ⎭ whichever comes first left to right

Example 5: Simplify each of the following.

(a) $16 + 4 \times 3$ (b) $10 - 3 \times 6 + 10^2 + (6 + 1) \times 4$

(a) First, the multiplication,

$$16 + 4 \times 3 = 16 + 12$$

Then, the addition,

$$16 + 12 = 28$$

(b) First, the operation inside the parentheses,

$$10 - 3 \times 6 + 10^2 + (6 + 1) \times 4 = 10 - 3 \times 6 + 10^2 + (7) \times 4$$

Then, the exponents ($10^2 = 10 \times 10 = 100$),

$$10 - 3 \times 6 + 10^2 + (7) \times 4 = 10 - 3 \times 6 + 100 + (7) \times 4$$

Then, multiplication,

$$10 - 3 \times 6 + 100 + (7) \times 4 = 10 - 18 + 100 + 28$$

Then, addition and subtraction left to right,

$$10 - 18 + 100 + 28 = -8 + 100 + 28$$
$$= 92 + 28$$
$$= 120$$

An easy way to remember the order of operations is: Please Excuse My Dear Aunt Sally (Parentheses, Exponents, Multiplication/ Division, Addition/Subtraction).

Place Value

Our number system is a **place value** system; that is, each place is assigned a different value. For instance, in the number 675, the 6 is in the hundreds place, the 7 is in the tens place, and the 5 is in the ones place. Since our number system is based on powers of 10 ($10^0 = 1$, $10^1 = 10$, $10^2 = 10 \times 10 = 100$, $10^3 = 10 \times 10 \times 10 = 1000$, and so on), each place is a progressive power of ten as follows.

billions	hundred millions	ten millions	millions	hundred thousands	ten thousands	thousands	hundreds	tens	ones
1,000,000,000	100,000,000	10,000,000	1,000,000	100,000	10,000	1000	100	10	1
10^9	10^8	10^7	10^6	10^5	10^4	10^3	10^2	10^1	10^0
							6	7	5

Notice how the number 675 fits into the place value grid.

Expanded Notation

Sometimes, numbers are written in **expanded notation** to point out the place value of each digit.

Example 1: Write 523 in expanded notation.

$$523 = 500 + 20 + 3$$
$$= (5 \times 100) + (2 \times 10) + (3 + 1)$$
$$= (5 \times 10^2) + (2 \times 10^1) + (3 \times 10^0)$$

These last two are the more common forms of expanded notation, one without exponents, one with exponents. Notice that in these the digit is multiplied times its place value: 1's, 10's, 100's, etc.

Example 2: Write 28,462 in expanded notation.

$$28{,}462 = 20{,}000 + 8000 + 400 + 60 + 2$$
$$= (2 \times 10{,}000) + (8 \times 1000) + (4 \times 100) + (6 \times 10)$$
$$+ (2 \times 1)$$
$$= (2 \times 10^4) + (8 \times 10^3) + (4 \times 10^2) + (6 \times 10^1)$$
$$+ (2 \times 10^0)$$

Rounding Off

To **round off** any number,

1. Underline the place value to which you're rounding off.

2. Look to the immediate right (one place) of your underlined place value.

3. Identify the number (the one to the right). If it is 5 or higher, round your underlined place value up 1 and change all the other numbers to its right to zeros. If the number (the one to the right) is 4 or less, leave your underlined place value as it is and change all the other numbers to its right to zeros.

Example 3: Round off 345,678 to the nearest thousand.

The number 345,678 is rounded up to 346,000.

Example 4: Round off 724,591 to the nearest ten thousand.

The number 724,591 is rounded down to 720,000.

Numbers that have been rounded off are called **rounded numbers.**

Estimating Sums, Differences, Products, and Quotients

Estimating sums. Use rounded numbers to estimate sums.

Example 5: Give an estimate for the sum 3741 + 5021 rounded to the nearest thousand.

$$3741 + 5021$$
$$\downarrow \qquad \downarrow$$
$$4000 + 5000 = 9000$$

So
$$3741 + 5021 \approx 9000$$

(The symbol $\approx$ means *is approximately equal to.*)

Estimating differences. Use rounded numbers to **estimate differences.**

Example 6: Give an estimate for the difference 317,753 − 115,522 rounded to the nearest hundred thousand.

$$317,753 - 115,522$$
$$\downarrow \qquad \downarrow$$
$$300,000 - 100,000 = 200,000$$

So $\qquad$ $317,753 - 115,222 \approx 200,000$

Estimating products. Use rounded numbers to **estimate products.**

Example 7: Estimate the product of 722 × 489 by rounding to the nearest hundred.

$$722 \times 489$$
$$\downarrow \qquad \downarrow$$
$$700 \times 500 = 350,000$$

So $\qquad$ $722 \times 489 \approx 350,000$

If both multipliers end in 50, or are halfway numbers, then rounding one number up and one number down will give you a better estimate of the product.

Example 8: Estimate the product of 650 × 350 by rounding to the nearest hundred.

$$650 \times 350$$
$$\downarrow \qquad \downarrow$$

Round one number up and one down. $\quad 700 \times 300 = 210{,}000$

So $\qquad\qquad\qquad\qquad\qquad\qquad 650 \times 350 \approx 210{,}000$

You could also round the first number down and the second number up and get this estimate.

$$650 \times 350$$
$$\downarrow \qquad \downarrow$$
$$600 \times 400 = 240{,}000$$

So $\qquad\qquad\qquad\qquad\qquad\qquad 650 \times 350 \approx 240{,}000$

In either case, your approximation will be closer than it would be if you rounded both numbers up, which is the standard rule.

Estimating quotients. Use rounded numbers to **estimate quotients.**

Example 9: Estimate the quotient of $891 \div 288$ by rounding to the nearest hundred.

$$891 \div 288$$
$$\downarrow \qquad \downarrow$$
$$900 \div 300 = 3$$

So $\qquad\qquad\qquad\qquad\qquad\qquad 891 \div 288 \approx 3$

Divisibility Rules

The following set of rules can help you save time in trying to check the divisibility of numbers. Instead of actually dividing into the numbers, try these rules.

A number is divisible by	if
2	it ends in 0, 2, 4, 6, or 8
3	the sum of its digits is divisible by 3
4	the number formed by the last two digits is divisible by 4
5	it ends in 0 or 5
6	it is divisible by 2 and 3 (use the rules for both)
7	(no simple rule)
8	The number formed by the last three digits is divisible by 8
9	the sum of its digits is divisible by 9

Example 10: Answer the following.

   (a) Is 126 divisible by 3?

   (b) Is 1648 divisible by 4?

   (c) Is 186 divisible by 6?

   (d) Is 2488 divisible by 8?

   (e) Is 2853 divisible by 8?

   (f) 4620 is divisible by which of the following numbers? 2, 3, 4, 5, 6, 7, 8, 9

   (a) Sum of digits of 126 = 9. Since 9 is divisible by 3, then 126 is divisible by 3.

   (b) Since 48 is divisible by 4, then 1648 is divisible by 4.

   (c) Since 186 ends in 6, it is divisible by 2. Sum of digits = 15. Since 15 is divisible by 3, then 186 is divisible by 3. 186 is divisible by 2 and 3; therefore, it is divisible by 6.

(d) Since 488 is divisible by 8, then 2488 is divisible by 8.

(e) Sum of digits of 2853 = 18. Since 18 is divisible by 9, then 2853 is divisible by 9.

(f) 4620 *is* divisible by

2—the number is even

3—the sum of the digits is 12, which is divible by 3

4—the number formed by the last two digits, 20, is divisible by 4

5—the number ends in 0

6—the number is divisible by 2 and 3

7—divide 4620 by 7 and you get 660

4620 is *not* divisible by

8—the number formed by the last three digits, 620, is not divisible by 8

9—the sum of digits is 12, which is not divisible by 9

Factors, Primes, Composites, and Factor Trees

Factors. Numbers that are multiplied together to give a product are called **factors.**

Example 11: What are the factors of 18?

$$\text{factor} \times \text{factor} = \text{product}$$
$$1 \times 18 = 18$$
$$2 \times 9 = 18$$
$$3 \times 6 = 18$$

So the factors of 18 are 1, 2, 3, 6, 9, and 18. These numbers are also called the **divisors** of 18. *Factors* of a number are also called *divisors* of that same number.

Prime numbers. A **prime number** is a number that can be divided by only itself and 1. Another definition: A prime number is a positive number that has exactly two different factors.

Example 12: Is 19 a prime number?

Yes. Since the only factors of 19 are 1 and 19, then 19 is a prime number. That is, since 19 is divisible by only 1 and 19, it is prime.

Example 13: Is 27 a prime number?

No. Since 27 is divisible by other numbers (3 and 9), then it is not prime. The factors of 27 are 1, 3, 9, and 27, so it is not prime.

The only even prime number is 2; thereafter, any even number may be divided by 2. The numbers 0 and 1 are not prime numbers. The prime numbers less than 50 are 2, 3, 5, 7, 11, 13, 17, 19, 23, 29, 31, 37, 41, 43, and 47.

Composite numbers. A **composite number** is a number divisible by more than just 1 and itself. Another definition: A composite number is a positive number that has more than two different factors. The numbers 4, 6, 8, 9, 10, 12, 14, 15, 16, 18, . . . are composite numbers, since they are "composed" of other numbers. The numbers 0 and 1 are not composite numbers (they are neither prime nor composite).

Example 14: Is 25 a composite number?

Yes. Since 25 is divisible by 5, it is composite. The factors of 25 are 1, 5, and 25.

Factor trees. Every composite number can be expressed as a product of **prime factors.** You can find prime factors by using a factor tree. A factor tree looks like this.

You could also make the tree as follows.

In either case, no matter how 18 is factored, the product of the primes is the same, even though the order may be different.

Example 15: Use a factor tree to express 60 as a product of prime factors.

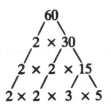

So the **prime factorization** of 60 is $2 \times 2 \times 3 \times 5$, which can be written as $2^2 \times 3 \times 5$. The actual *prime factors* of 60 are 2, 3, and 5.

Definition of the Decimal System

The system of numbers that you use is called the **decimal system** and, as mentioned earlier, is based on powers of ten (**base ten system**). Each place in the place value grid is ten times the value of the place to the right of it. Notice this version of the place value grid shown earlier. Every number to the right of the decimal point is a **decimal fraction** (a fraction with a denominator of 10, 100, 1000, etc.)

millions	hundred thousands	ten thousands	thousands	hundreds	tens	ones	tenths	hundredths	thousandths	ten thousandths	hundred thousandths
							$1/10$	$1/100$	$1/1000$	$1/10,000$	$1/100,000$
1,000,000	100,000	10,000	1000	100	10		.1	.01	.001	.0001	.00001
10^6	10^5	10^4	10^3	10^2	10^1	10^0	10^{-1}	10^{-2}	10^{-3}	10^{-4}	10^{-5}

Notice that $\frac{1}{10}$ can be written as ten to a negative exponent, 10^{-1}. Similarly, $\frac{1}{100}$ can be written as ten to a negative exponent, 10^{-2}. Negative exponents will be discussed in a later section.

Using the Place Value Grid

Expanded notation. Decimals can also be written in expanded notation, using the same techniques as when expanding whole numbers.

Example 1: Write .365 in expanded notation.

$$.365 = .3 + .06 + .005$$
$$= (3 \times .1) + (6 \times .01) + (5 \times .001)$$
$$= (3 \times 10^{-1}) + (6 \times 10^{-2}) + (5 \times 10^{-3})$$

Example 2: Write 5.26 in expanded notation.

$$5.26 = 5 + .2 + .06$$
$$= (5 \times 1) + (2 \times .1) + (6 \times .01)$$
$$= (5 \times 10^0) + (2 \times 10^{-1}) + (6 \times 10^{-2})$$

Writing decimals. To **read a decimal** or **write a decimal in words,** you start at the left and end with the place value of the last number on the right. When there is a whole number included, use the word *and* to show the position of the decimal point.

Example 3: Read the number .75.

seventy-five hundredths

Example 4: Read the number 45.321.

forty-five *and* three hundred twenty-one thousandths

Example 5: Write two hundred *and* three tenths.

200.3

Comparing decimals. If you want to **compare decimals,** that is, find out if one decimal is greater than another, simply make sure that each decimal goes out to the same number of places to the right.

Example 6: Which is greater, .37 or .365?

Since .37 = .370, you can now align the two decimals.

.370
.365

It is easy to see that .37 is greater. Here, you are really comparing three hundred seventy thousandths to three hundred sixty-five thousandths.

Example 7: Put the decimals .66, .6587, and .661 in order from largest to smallest.

First, change each number to ten-thousandths by adding zeros where appropriate. Then, align the decimal points to make the comparison.

$$.6600$$
$$.6587$$
$$.6610$$

The order should be .661, .66, and .6587.

You might also align the decimals first and then add the zeros as follows.

$$.66 \quad = .6600$$
$$.6587 = .6587$$
$$.661 \quad = .6610$$

Remember, the number of digits to the right of the decimal point does not determine the size of the number (.5 is greater than .33).

Rounding Decimals

The method for rounding decimals is almost identical to the method used for rounding whole numbers. To round off a decimal,

1. Underline the place value to which you're rounding.

2. Look to the immediate right (one place) of your underlined place value.

3. Identify the number (the one to the right). If it is 5 or higher, round your underlined place value up 1 and *drop all the numbers to the right of your underlined number.* If the number

(the one to the right) is 4 or less, leave your underlined place value as it is and *drop all the numbers to the right of your underlined number.*

Example 8: Round off .478 to the nearest hundredth.

.47̲8 is rounded up to .48

Example 9: Round off 5.3743 to the nearest thousandth.

5.374̲3 is rounded down to 5.374

Decimal Computation

Adding and subtracting decimals. To **add** or **subtract decimals,** just line up the decimal points and then add or subtract in the same manner you would add or subtract whole numbers.

Example 10: Add 23.6 + 1.75 + 300.002.

$$
\begin{array}{r}
23.6 \\
1.75 \\
300.002 \\
\hline
325.352
\end{array}
$$

Adding in zeros can make the problem easier to work.

$$
\begin{array}{r}
23.600 \\
1.750 \\
300.002 \\
\hline
325.352
\end{array}
$$

Example 11: Subtract 54.26 − 1.1.

$$
\begin{array}{r}
54.26 \\
-\ 1.10 \\
\hline
53.16
\end{array}
$$

Example 12: Subtract 78.9 − 37.43.

$$
\begin{array}{r}
\overset{8}{78.\overset{1}{\cancel{9}}0} \\
-37.4\,3 \\
\hline
41.4\,7
\end{array}
$$

A whole number has an understood decimal point to its right.

Example 13: Subtract 17 − 8.43.

$$
\begin{array}{r}
\overset{6\ 9}{17.\overset{1}{\cancel{0}}0} \\
-\ 8.4\,3 \\
\hline
8.5\,7
\end{array}
$$

Multiplying decimals. To **multiply decimals,** just multiply as usual. Then count the total number of digits above the line which are to the right of all decimal points. Place your decimal point in your answer so there is the same number of digits to the right of it as there was above the line.

Example 14: Multiply 40.012 × 3.1.

```
  40.012 ← 3 digits        { total of 4 digits above the line
×    3.1 ← 1 digit         { that are to the right of the decimal point
  40 012
1200 36
─────────
124.0372 ← 4 digits        { decimal point placed so there is
                           { same number of digits to the right
                           { of the decimal point
```

Dividing decimals. Dividing decimals is the same as dividing other numbers, except that if the divisor (the number you're dividing by) has a decimal, move it to the right as many places as necessary until it is a whole number. Then, move the decimal point in the dividend (the number being divided into) the same number of places. Sometimes, you may have to add zeros to the dividend (the number inside the division sign).

Example 15: Divide 1.25$\overline{)5}$.

$$125\overline{)5.} = 125\overline{)500.}^{\;\;\;4.}$$

Example 16: Divide 0.002$\overline{)26}$.

$$0.002\overline{)26.} = 2\overline{)26000.}^{\;\;\;13000.}$$

Example 17: Divide $20\overline{)13}$.

$$20\overline{)13.} = 20\overline{)\begin{array}{r} .65 \\ \hline 13.00 \\ 12\ 0 \\ \hline 1\ 00 \\ 1\ 00 \\ \hline 0 \end{array}}$$

Estimating Sums, Differences, Products, and Quotients

When working with decimals, it is easy to make a simple mistake and misplace the decimal point. Estimating an answer can be a valuable tool in helping you avoid this type of mistake.

Estimating sums. Use rounded numbers to **estimate sums.**

Example 18: Give an estimate for the sum of 19.61 and 5.07 by rounding to the nearest tenth.

Round each number to the nearest tenth.

$$19.61 \quad + 5.07$$
$$\downarrow \qquad \downarrow$$
$$19.6 \quad + 5.1$$

$$\begin{array}{r} 19.6 \\ + \ 5.1 \\ \hline 24.7 \end{array}$$

So $\qquad 19.61 + 5.07 \approx 24.7$

Example 19: Estimate the sum of 19.61 + 5.07 by rounding to the nearest whole number.

Round each number to a whole number.

$$19.61 + 5.07$$
$$\downarrow \qquad \downarrow$$
$$20 \quad + \ 5 \ = 25$$

So $\qquad 19.61 + 5.07 \approx 25$

Estimating differences. Use rounded numbers to **estimate differences.**

Example 20: Give an estimate for the difference of 12.356 − 5.281 by rounding to the nearest whole number.

Round each number to the nearest whole number.

$$12.356 - 5.281$$
$$\downarrow \qquad \downarrow$$
$$12 \quad - \ 5$$

Now, subtract.

$$\begin{array}{r} 12 \\ - \ 5 \\ \hline 7 \end{array}$$

So $\qquad 12.356 - 5.281 \approx 7$

Estimating products. Use rounded numbers to **estimate products.**

Example 21: Estimate the product of 4.7 × 5.9 by rounding to the nearest whole number.

Round each number to a whole number.

$$4.7 \times 5.9$$
$$\downarrow \quad \downarrow$$
$$5 \ \times 6 \ = 30$$

So $\qquad 4.7 \times 5.9 \approx 30$

Again in decimals, as in whole numbers, if both multipliers end in .5, or are halfway numbers, then rounding one number up and one number down will give you a better estimate of the product.

Example 22: Estimate the product of 7.5 × 8.5 by rounding to the nearest whole number.

$$7.5 \times 8.5$$
$$\downarrow \quad \downarrow$$
Round one number up and one down. $\quad 8 \ \times 8 \ = 64$

So $\qquad 7.5 \times 8.5 \approx 64$

You could also round the first number down and the second number up and get this estimate.

$$7.5 \times 8.5$$
$$\downarrow \quad \downarrow$$
$$7 \ \times 9 \ = 63$$

So $\qquad 7.5 \times 8.5 \approx 63$

In either case, your approximation will be closer than it would be if you rounded both numbers up, which is the standard rule.

Estimating quotients. Use rounded numbers to **estimate quotients.**

Example 23: Estimate the quotient of 27.49 ÷ 3.12 by rounding to the nearest whole number.

Round each number to the nearest whole number.

$$27.49 \div 3.12$$
$$\downarrow \qquad \downarrow$$
$$27 \quad \div 3 \quad = 9$$

So $\qquad\qquad$ 27.49 ÷ 3.12 ≈ 9

Repeating Decimals

The most commonly used decimals are **terminating decimals,** that is, decimals that stop, such as .5 or .74. A **repeating decimal** is a decimal that continues on indefinitely and repeats a number or block of numbers in a consistent manner, such as .666 . . . or .232323 . . . A **vinculum** (a horizontal line over the number or numbers) is the standard notation used to show that a number or group of numbers is repeating. Using the vinculum, the repeating decimal looks like this: $.\overline{6}$ or $.\overline{23}$. Some books put the vinculum below the number, but this is less common.

A **fraction,** or **fractional number,** is used to represent a part of a whole. Fractions consist of two numbers: a **numerator** (which is above the line) and a **denominator** (which is below the line).

$$\frac{1}{2} \quad \begin{matrix} \text{numerator} \\ \text{denominator} \end{matrix}$$

or

$$1/2 \quad \text{numerator/denominator}$$

The denominator lets us know the number of equal parts into which something is divided. The numerator tells us how many of these equal parts are being considered. Thus, if the fraction is $\frac{3}{5}$ of a pie, then the denominator 5 tells us that the pie has been divided into 5 equal parts, of which 3 (numerator) are in the fraction. Sometimes, it helps to think of the dividing line (in the middle of a fraction) as meaning "out of." In other words, $\frac{3}{5}$ would also mean 3 "out of" 5 equal pieces from the whole pie.

Proper and Improper Fractions

A fraction like $\frac{3}{5}$, where the numerator is smaller than the denominator, is less than one. This kind of fraction is called a **proper fraction.** But sometimes a fraction may be more than one. This is when the numerator is larger than the denominator. Thus, $\frac{12}{7}$ is more than one. This kind of fraction is called an **improper fraction.**

Examples of proper fractions: $\frac{4}{7}, \frac{2}{5}, \frac{1}{9}, \frac{10}{12}$

Examples of improper fractions: $\frac{7}{4}, \frac{3}{2}, \frac{10}{3}, \frac{16}{15}$

Mixed Numbers

When a term contains both a whole number (3, 8, 25, etc.) and a fraction ($\frac{1}{2}$, $\frac{1}{4}$, $\frac{3}{4}$, etc.), it is called a **mixed number.** For instance, $5\frac{1}{4}$ and $290\frac{3}{4}$ are both mixed numbers.

Changing improper fractions to mixed numbers. To change an improper fraction to a mixed number, you divide the denominator into the numerator.

Example 1: Change $\frac{10}{3}$ to a mixed number.

$$
\begin{array}{r}
3 \\
3 \overline{)10} \\
9 \\
\hline
1 \quad \text{\scriptsize remainder}
\end{array}
$$

$$\frac{10}{3} = 3\frac{1}{3}$$

Changing mixed numbers to improper fractions. To change a mixed number to an improper fraction, you multiply the denominator times the whole number, add in the numerator, and put the total over the original denominator.

Example 2: Change $5\frac{3}{4}$ to an improper fraction.

$$4 \times 5 + 3 = 23$$

$$5\frac{3}{4} = \frac{23}{4}$$

Renaming Fractions

Equivalent fractions. Fractions that name the same number, such as $\frac{1}{2}$, $\frac{2}{4}$, $\frac{3}{6}$, $\frac{4}{8}$, and $\frac{5}{10}$, are called **equivalent fractions.** A simple method to check if fractions are equivalent is to cross multiply and check the products.

Example 3: Is $\frac{2}{4}$ equivalent to $\frac{3}{6}$?

$$12 = 12$$
$$\frac{2}{4} \bowtie \frac{3}{6}$$

Since the cross products are the same, the fractions $\frac{2}{4}$ and $\frac{3}{6}$ are equivalent.

Example 4: Is $\frac{3}{4}$ equivalent to $\frac{2}{3}$?

$$9 \neq 8$$
$$\frac{3}{4} \bowtie \frac{2}{3}$$

Since the cross products are not the same, the fractions $\frac{3}{4}$ and $\frac{2}{3}$ are not equivalent.

Reducing fractions. When given as a final answer, a fraction should be reduced to lowest terms. **Reducing fractions** is done by dividing both the numerator and denominator by the largest number that will divide evenly into both.

Example 5: Reduce $\frac{15}{25}$ to lowest terms.

To reduce $\frac{15}{25}$ to lowest terms, divide the numerator and denominator by 5.

$$\frac{15}{25} = \frac{15 \div 5}{25 \div 5} = \frac{3}{5}$$

Since $\frac{3}{5}$ cannot be reduced any further, that is, the numerator and denominator cannot both be evenly divided again, $\frac{3}{5}$ is reduced to **lowest terms.**

Example 6: Reduce $\frac{8}{40}$ to lowest terms.

$$\frac{8}{40} = \frac{8 \div 8}{40 \div 8} = \frac{1}{5}$$

Enlarging denominators. The **denominator** of a fraction may be **enlarged** by multiplying both the numerator and denominator by the same number.

Example 7: Change $\frac{3}{4}$ to eighths.

To change $\frac{3}{4}$ to eighths, simply multiply the numerator and denominator by 2

$$\frac{3}{4} = \frac{3 \times 2}{4 \times 2} = \frac{6}{8}$$

Example 8: Express $\frac{1}{2}$ as tenths.

$$\frac{1}{2} = \frac{1 \times 5}{2 \times 5} = \frac{5}{10}$$

Factors

As mentioned earlier, **factors** of a number are those whole numbers which when multiplied together yield the number.

Example 9: What are the factors of 10?

Since $\qquad 10 = 2 \times 5$

and $\qquad 10 = 1 \times 10$

the factors of 10 are 1, 2, 5, and 10.

Example 10: What are the factors of 24?

$$24 = 1 \times 24$$
$$= 2 \times 12$$
$$= 3 \times 8$$
$$= 4 \times 6$$

Therefore, the factors of 24 are 1, 2, 3, 4, 6, 8, 12, and 24.

Common factors. Common factors are those factors that are the same for two or more numbers.

Example 11: What are the common factors of 6 and 8?

Factors of 6:   ①② 3 6

Factors of 8:   ①② 4 8

1 and 2 are common factors of 6 and 8.

Note that some numbers may have many common factors.

Example 12: What are the common factors of 24 and 36?

Factors of 24:   ①②③④⑥ 8 ⑫ 24

Factors of 36:   ①②③④⑥ 9 ⑫ 18 36

Thus, the common factors of 24 and 36 are 1, 2, 3, 4, 6, and 12.

Greatest common factor. The **greatest common factor (GCF)** is the largest factor common to two or more numbers.

Example 13: What is the greatest common factor of 12 and 30?

Factors of 12:   ①②③ 4 ⑥ 12

Factors of 30:   ①②③ 5 ⑥ 10 15 30

Notice that, while 1, 2, 3, and 6 are all common factors of 12 and 30, only 6 is the greatest common factor.

Multiples

Multiples of a number are found by multiplying that number by 1, by 2, by 3, by 4, by 5, etc.

Example 14: List the first seven multiples of each number.

    (a) 2   (b) 5   (c) 9

    (a)   2, 4, 6, 8, 10, 12, 14

    (b)   5, 10, 15, 20 25, 30, 35

    (c)   9, 18, 27, 36, 45, 54, 63

Common multiples. Common multiples are those multiples that are the same for two or more numbers.

Example 15: What are the common multiples of 2 and 3?

Multiples of 2:   2   4   ⑥   8   10   ⑫   14   16   ⑱   etc.
Multiples of 3:   3   ⑥   9   ⑫   15   ⑱   etc.

Notice that common multiples may go on indefinitely.

Least common multiple. The **least common multiple (LCM)** is the smallest multiple that is common to two or more numbers.

Example 16: What is the least common multiple of 2 and 3?

Multiples of 2:   2   4   ⑥   8   10   ⑫   etc.

Multiples of 3:   3   ⑥   9   ⑫   etc.

The smallest multiple common to both 2 and 3 is 6.

Example 17: What is the least common multiple of 2, 3, and 4?

Multiples of 2:   2   4   6   8   10   ⑫   etc.

Multiples of 3:   3   6   9   ⑫   etc.

Multiples of 4:   4   8   ⑫   etc.

The least common multiple of 2, 3, and 4 is 12.

Adding and Subtracting Fractions

Adding fractions. To **add fractions,** you must have a common denominator. Fractions that have common denominators are called **like fractions.** Fractions that have different denominators are called **unlike fractions.** To add *like* fractions, simply add the numerators and keep the same (or like) denominator.

Example 18: Add $\frac{1}{5} + \frac{3}{5}$.

$$\begin{array}{r} \frac{1}{5} \\ + \frac{3}{5} \\ \hline \frac{4}{5} \end{array}$$

To add *unlike* fractions, first change all denominators to their **lowest common denominator (LCD),** also called the lowest common multiple of the denominator, the lowest number than can be divided evenly by all denominators in the problem. The numerators may need to be changed to make sure that the fractions are still equivalent to the originals. When you have all the denominators the same, you may add the numerators and keep the same denominator.

Example 19: Add

(a) $\frac{3}{8} + \frac{1}{2}$ (b) $\frac{1}{4} + \frac{1}{3}$

(a) Change the $\frac{1}{2}$ to $\frac{4}{8}$ because 8 is the lowest common denominator, and then add the numerators 3 and 4 to get $\frac{7}{8}$.

$$\begin{array}{r} \frac{3}{8} = \frac{3}{8} \\ + \frac{1}{2} = \frac{4}{8} \leftarrow \text{(change } \frac{1}{2} \text{ to } \frac{4}{8}) \\ \hline \frac{7}{8} \end{array}$$

(b) Change both fractions to get the lowest common denominator of 12, and then add the numerators to get $\frac{7}{12}$.

$$\begin{array}{r} \frac{1}{4} = \frac{3}{12} \\ + \frac{1}{3} = \frac{4}{12} \swarrow \text{(change both fractions to LCD of 12)} \\ \hline \frac{7}{12} \end{array}$$

Note that fractions may be added across as well.

Example 20: Add $\frac{1}{2} + \frac{1}{3}$.

$$\frac{1}{2} + \frac{1}{3} = \frac{3}{6} + \frac{2}{6} = \frac{5}{6}$$

Subtracting fractions. To **subtract fractions,** the same rule as in adding fractions (find the LCD) applies, except that you subtract the numerators.

Example 21: Subtract

(a) $\frac{7}{8} - \frac{1}{4}$ (b) $\frac{3}{4} - \frac{1}{3}$

(a)
$$\begin{aligned}\frac{7}{8} &= \frac{7}{8} \\ -\frac{1}{4} &= \frac{2}{8} \\ \hline &\frac{5}{8}\end{aligned}$$

(b)
$$\begin{aligned}\frac{3}{4} &= \frac{9}{12} \\ -\frac{1}{3} &= \frac{4}{12} \\ \hline &\frac{5}{12}\end{aligned}$$

Again, a subtraction problem may be done across as well as down

Example 22: Subtract $\frac{5}{8} - \frac{3}{8}$.

$$\frac{5}{8} \quad \frac{3}{8} = \frac{2}{8} = \frac{1}{4}$$

Adding and Subtracting Mixed Numbers

Adding mixed number. To **add mixed numbers,** the same rule as in adding fractions (find the LCD) applies, but make sure that you always add the *whole numbers* to get your final answer.

Example 23: Add $2\frac{1}{2} + 3\frac{1}{4}$.

$$2\frac{1}{2} = 2\frac{2}{4} \leftarrow (\frac{1}{2} \text{ is changed to } \frac{2}{4})$$
$$+ 3\frac{1}{4} = 3\frac{1}{4}$$
$$5\frac{3}{4}$$

↑
(remember to add the whole numbers)

Sometimes, you may end up with a mixed number that includes an improper fraction. In that case, you *must* change the improper fraction to a mixed number and combine it with the sum of the integers.

Example 24: Add $2\frac{1}{2} + 5\frac{3}{4}$.

$$2\frac{1}{2} = 2\frac{2}{4}$$
$$+ 5\frac{3}{4} = 5\frac{3}{4}$$
$$7\frac{5}{4}$$

And since $\frac{5}{4} = 1\frac{1}{4}$,

$$7\frac{5}{4} = 7 + 1\frac{1}{4} = 8\frac{1}{4}$$

Subtracting mixed numbers. When you subtract mixed numbers, you sometimes may have to "borrow" from the whole number, just as you sometimes borrow from the next column when subtracting whole numbers. Note that when you borrow 1 from the whole number, the 1 must be changed to a fraction.

Example 25: Subtract

(a) $651 - 129$ (b) $4\frac{1}{6} - 2\frac{5}{6}$ (c) $5\frac{1}{5} - 3\frac{1}{2}$

(a)
$$\begin{array}{r} \overset{4\ 11}{6\cancel{5}1} \\ -\ 129 \\ \hline 522 \end{array}$$

(you borrowed 1 from the 10's column)

(b)
$$\begin{array}{r} 3\frac{7}{6} \\ \cancel{4}\cancel{\frac{1}{6}} \\ -\ 2\frac{5}{6} \\ \hline 1\frac{2}{6} = 1\frac{1}{3} \end{array}$$

(you borrowed 1 in the form $\frac{6}{6}$ from the 1's column)

(c)
$$\begin{array}{r} 5\frac{1}{5} = 5\frac{2}{10} = 4\frac{12}{10} \\ -\ 3\frac{1}{2} = 3\frac{5}{10} = 3\frac{5}{10} \\ \hline 1\frac{7}{10} \end{array}$$

Notice that you should borrow only after you have gotten a common denominator.

To subtract a mixed number from a whole number, you have to "borrow" from the whole number.

Example 26: Subtract $6 - 3\frac{1}{5}$.

$$\begin{array}{r} 6\ =\ 5\frac{5}{5} \leftarrow \text{(borrow 1 in the form of } \frac{5}{5} \text{ from the 6)} \\ -\ 3\frac{1}{5} =\ 3\frac{1}{5} \\ \hline 2\frac{4}{5} \\ \uparrow \end{array}$$

(remember to subtract the remaining whole numbers)

Multiplying Fractions and Mixed Numbers

Multiplying fractions. To **multiply fractions,** simply multiply the numerators; then multiply the denominators. Reduce to lowest terms if necessary.

Example 27: Multiply $\frac{2}{3} \times \frac{5}{12}$.

$$\frac{2}{3} \times \frac{5}{12} = \frac{10}{36}$$

Reduce.

$$\frac{10}{36} = \frac{5}{18}$$

This answer had to be reduced because it wasn't in lowest terms.

You could first **cancel** when multiplying fractions, which eliminates the need to reduce your answer. To cancel, find a number that divides evenly into one numerator and one denominator. In Example 27, 2 will divide evenly into 2 in the numerator (it goes in one time) and 12 in the denominator (it goes in six times). Thus,

$$\frac{\overset{1}{\cancel{2}}}{3} \times \frac{5}{\underset{6}{\cancel{12}}} = \frac{5}{18}$$

Since whole numbers can also be written as fractions ($3 = \frac{3}{1}$, $4 = \frac{4}{1}$, etc.), the problem $3 \times \frac{3}{8}$ would be worked by changing 3 to $\frac{3}{1}$.

Example 28: Multiply $3 \times \frac{3}{8}$.

$$3 \times \frac{3}{8} = \frac{3}{1} \times \frac{3}{8}$$

$$= \frac{9}{8}$$

$$= 1\frac{1}{8}$$

Example 29: Multiply $\frac{1}{4} \times \frac{2}{7}$.

$$\frac{1}{4} \times \frac{2}{7} = \frac{1}{\overset{}{\underset{2}{\cancel{4}}}} \times \frac{\overset{1}{\cancel{2}}}{7} = \frac{1}{14}$$

Remember, you may cancel *only* when *multiplying* fractions.

Multiplying mixed numbers. To **multiply mixed numbers,** first change any mixed number to an improper fraction. Then multiply the numerators together and the denominators together, as previously shown.

Example 30: Multiply $3\frac{1}{3} \times 2\frac{1}{4}$.

$$3\tfrac{1}{3} \times 2\tfrac{1}{4} = \tfrac{10}{3} \times \tfrac{9}{4} = \tfrac{90}{12} = 7\tfrac{6}{12} = 7\tfrac{1}{2}$$

Or

$$\frac{\overset{5}{\cancel{10}}}{\underset{1}{\cancel{3}}} \times \frac{\overset{3}{\cancel{9}}}{\underset{2}{\cancel{4}}} = \frac{15}{2} = 7\tfrac{1}{2}$$

Example 31: Multiply $3\frac{1}{5} \times 6\frac{1}{2}$.

$$3\tfrac{1}{5} \times 6\tfrac{1}{2} = \frac{16}{5} \times \frac{13}{2} = \frac{\overset{8}{\cancel{16}}}{5} \times \frac{13}{\underset{1}{\cancel{2}}} = \frac{104}{5} = 20\tfrac{4}{5}$$

Dividing Fractions and Mixed Numbers

Dividing fractions. To **divide fractions,** invert (turn upside down) the second fraction (the one "divided by") and multiply. Then reduce, if necessary.

Example 32: Divide

(a) $\frac{1}{6} \div \frac{1}{5}$ (b) $\frac{1}{6} \div \frac{1}{3}$

(a)
$$\frac{1}{6} \div \frac{1}{5} = \frac{1}{6} \times \frac{5}{1} = \frac{5}{6}$$

(b)
$$\frac{1}{6} \div \frac{1}{3} = \frac{1}{\cancel{6}_2} \times \frac{\cancel{3}^1}{1} = \frac{1}{2}$$

Example 33: Divide $\frac{1}{3} \div 6$.

Since $6 = \frac{6}{1}$, the problem can be written $\frac{1}{3} \div \frac{6}{1}$. Then invert the second fraction and multiply.

$$\frac{1}{3} \div \frac{6}{1} = \frac{1}{3} \times \frac{1}{6} = \frac{1}{18}$$

From Example 33, since 6 and $\frac{1}{6}$ are reciprocals, you can see that dividing is the same as multiplying by the reciprocal

Example 34: Divide $\frac{3}{7} \div \frac{3}{14}$.

$$\frac{3}{7} \div \frac{3}{14} = \frac{\cancel{3}}{\cancel{7}_1} \times \frac{\cancel{14}^2}{\cancel{3}} = \frac{2}{1} = 2$$

Dividing complex fractions. Sometimes, a division of fractions problem may appear in this form (these are called **complex fractions**).

$$\frac{\frac{3}{4}}{\frac{7}{8}}$$

If so, consider the line separating the two fractions to mean "divided by." Therefore, this problem may be rewritten as

$$\tfrac{3}{4} \div \tfrac{7}{8}$$

Now, follow the same procedure as shown on page 47.

$$\frac{3}{4} \div \frac{7}{8} = \frac{3}{\cancel{4}_1} \times \frac{\cancel{8}^2}{7} = \frac{6}{7}$$

Example 35: Divide $\dfrac{\frac{7}{8}}{\frac{1}{2}}$.

$$\frac{\frac{7}{8}}{\frac{1}{2}} = \frac{7}{8} \div \frac{1}{2} = \frac{7}{\cancel{8}_4} \times \frac{\cancel{2}^1}{1} = \frac{7}{4} = 1\tfrac{3}{4}$$

Dividing mixed numbers. To **divide mixed numbers,** first change them to improper fractions (page 34). Then follow the rule for dividing fractions (page 47).

Example 36: Divide

(a) $3\tfrac{3}{5} \div 2\tfrac{2}{3}$ (b) $2\tfrac{1}{5} \div 3\tfrac{1}{10}$

(a) $$3\tfrac{3}{5} \div 2\tfrac{2}{3} = \frac{18}{5} \div \frac{8}{3} = \frac{\cancel{18}^9}{5} \times \frac{3}{\cancel{8}_4} = \frac{27}{20} = 1\tfrac{7}{20}$$

(b) $$2\tfrac{1}{5} \div 3\tfrac{1}{10} = \frac{11}{5} \div \frac{31}{10} = \frac{11}{\cancel{5}_1} \times \frac{\cancel{10}^2}{31} = \frac{22}{31}$$

Notice that after you invert and have a multiplication of fractions problem, you may then cancel tops with bottoms when appropriate

Simplifying Fractions and Complex Fractions

If either numerator or denominator consists of several numbers, these numbers must be combined into one number. Then reduce if necessary.

Example 37: Simplify

(a) $\dfrac{28 + 14}{26 + 17}$ (c) $\dfrac{2 + \frac{1}{2}}{3 + \frac{1}{4}}$ (e) $\dfrac{1}{1 + \dfrac{1}{1 + \frac{1}{4}}}$

(b) $\dfrac{\frac{1}{4} + \frac{1}{2}}{\frac{1}{3} + \frac{1}{4}}$ (d) $\dfrac{3 - \frac{3}{4}}{4 + \frac{1}{2}}$

(a)

$$\frac{28 + 14}{26 - 17} = \frac{42}{43}$$

(b)

$$\frac{\frac{1}{4} + \frac{1}{2}}{\frac{1}{3} + \frac{1}{4}} = \frac{\frac{1}{4} + \frac{2}{4}}{\frac{4}{12} + \frac{3}{12}}$$

$$= \frac{\frac{3}{4}}{\frac{7}{12}}$$

$$= \frac{3}{4} \div \frac{7}{12}$$

$$= \frac{3}{\underset{1}{\cancel{4}}} \times \frac{\overset{3}{\cancel{12}}}{7}$$

$$= \frac{9}{7}$$

$$= 1\frac{2}{7}$$

(c)

$$\frac{2 + \frac{1}{2}}{3 + \frac{1}{4}} = \frac{2\frac{1}{2}}{3\frac{1}{4}}$$

$$= \frac{\frac{5}{2}}{\frac{13}{4}}$$

$$= \frac{5}{2} \div \frac{13}{4}$$

$$= \frac{5}{\cancel{2}_1} \times \frac{\cancel{4}^2}{13}$$

$$= \frac{10}{13}$$

(d)

$$\frac{3 - \frac{3}{4}}{4 + \frac{1}{2}} = \frac{2\frac{1}{4}}{4\frac{1}{2}}$$

$$= \frac{\frac{9}{4}}{\frac{9}{2}}$$

$$= \frac{9}{4} \div \frac{9}{2}$$

$$= \frac{\cancel{9}}{\cancel{4}_2} \times \frac{\cancel{2}^1}{\cancel{9}}$$

$$= \frac{1}{2}$$

(e)
$$\cfrac{1}{1 + \cfrac{1}{1 + \frac{1}{4}}} = \cfrac{1}{1 + \cfrac{1}{\frac{5}{4}}}$$

$$= \cfrac{1}{1 + (1 \div \frac{5}{4})}$$

$$= \cfrac{1}{1 + (1 \times \frac{4}{5})}$$

$$= \cfrac{1}{1 + \frac{4}{5}}$$

$$= \cfrac{1}{1\frac{4}{5}}$$

$$= \cfrac{1}{\frac{9}{5}}$$

$$= 1 \div \frac{9}{5}$$

$$= 1 \times \frac{5}{9}$$

$$= \frac{5}{9}$$

Changing Fractions to Decimals

Fractions may also be written in **decimal** form (**decimal fractions**) as either **terminating** (for example, .3) or **infinite repeating** (for example, .66 . . .) decimals. To **change a fraction to a decimal,** simply do what the operation says. In other words, $\frac{13}{20}$ means 13 divided by 20. So do just that (insert decimal points and zeros accordingly).

Example 38: Change to decimals.

(a) $\frac{5}{8}$ (b) $\frac{2}{5}$ (c) $\frac{4}{9}$

(a)

$$\frac{5}{8} = 8\overline{)\begin{array}{l} .625 \\ 5.000 \end{array}}$$
$$\begin{array}{r} 48 \\ \hline 20 \\ 16 \\ \hline 40 \\ 40 \\ \hline 0 \end{array}$$

So $\frac{5}{8} = .625$.

(b)

$$\frac{2}{5} = 5\overline{)\begin{array}{l} .4 \\ 2.0 \end{array}}$$
$$\begin{array}{r} 20 \\ \hline 0 \end{array}$$

So $\frac{2}{5} = .4$

(c)

$$\frac{4}{9} = 9\overline{)\begin{array}{l} .444 .. \\ 4.000 \end{array}}$$
$$\begin{array}{r} 36 \\ \hline 40 \\ 36 \\ \hline 40 \end{array}$$

So $\frac{4}{9} = .444\ldots$ or $.\overline{4}$.

Changing Terminating Decimals to Fractions

To **change terminating decimals to fractions,** simply remember that all numbers to the right of the decimal point are fractions with denominators of only 10, 100, 1000, 10,000, etc. Next, use the technique of *read it, write it,* and *reduce it.*

Example 39: Change the following to fractions in lowest terms.

(a) .8   (b) .09

(a)   *Read it:*   .8   (eight tenths)

             Write it: $\frac{8}{10}$

             Reduce it: $\frac{4}{5}$

(b)   *Read it:*   .09   (nine hundredths)

             Write it: $\frac{9}{100}$

             Reduce it: $\frac{9}{100}$ can't be reduced

Changing Infinite Repeating Decimals to Fractions

Remember, infinite repeating decimals are usually represented by putting a line over (sometimes under) the shortest block of repeating decimals. Every **infinite repeating decimal can be expressed as a fraction.**

Example 40: Find the fraction represented by the repeating decimal $.\overline{7}$.

Let n stand for $\qquad$ $.\overline{7}$ or $.77777\ldots$

So $10n$ stands for $\qquad$ $7.\overline{7}$ or $7.77777\ldots$

Since $10n$ and n have the same fractional part, their difference is an integer.

$$
\begin{array}{r}
10n = 7.\overline{7} \\
-\quad n = \ .\overline{7} \\
\hline
9n = 7
\end{array}
$$

You can solve this problem as follows.

$$9n = 7$$

$$n = \tfrac{7}{9}$$

So $\qquad$ $.\overline{7} = \tfrac{7}{9}$

Example 41: Find the fraction represented by the repeating decimal $.\overline{36}$.

Let n stand for $\qquad$ $.\overline{36}$ or $.363636\ldots$

So $10n$ stands for $\qquad$ $3.6\overline{36}$ or $3.63636\ldots$

and $100n$ stands for $\qquad$ $36.\overline{36}$ or $36.3636\ldots$

Since $100n$ and n have the same fractional part, their difference is an integer. (The repeating parts are the same and subtract out.)

$$
\begin{array}{r}
100n = 36.\overline{36} \\
-\quad n = \quad .\overline{36} \\
\hline
99n = 36
\end{array}
$$

You can solve this equation as follows.

$$99n = 36$$

$$n = \tfrac{36}{99}$$

Now, reduce $\tfrac{36}{99}$ to $\tfrac{4}{11}$.

So $\qquad\qquad\qquad\qquad .\overline{36} = \tfrac{4}{11}$

Example 42: Find the fraction represented by the repeating decimal $.5\overline{4}$.

Let n stand for	$.5\overline{4}$ or	$.544444\ldots$
So $10n$ stands for	$5.\overline{4}$ or	$5.44444\ldots$
and $100n$ stands for	$54.\overline{4}$ or	$54.4444\ldots$

Since $100n$ and $10n$ have the same fractional part, their difference is an integer. (Again, notice how the repeated parts must align to subtract out.)

$$\begin{aligned} 100n &= 54.\overline{4} \\ -\ 10n &= \ \ 5.\overline{4} \\ \hline 90n &= 49 \end{aligned}$$

You can solve this equation as follows.

$$90n = 49$$

$$n = \tfrac{49}{90}$$

So $\qquad\qquad\qquad\qquad .5\overline{4} = \tfrac{49}{90}$

A fraction whose denominator is 100 is called a **percent**. The word *percent* means hundredths (per hundred). So $43\% = \frac{43}{100}$.

Changing Percents, Decimals, and Fractions

Changing decimals to percents. To **change decimals to percents,**

1. Move the decimal point two places to the right.
2. Insert a percent sign.

Example 1: Change to percents.

(a) .55   (b) .02   (c) 3.21   (d) .004

(a)   .55 = 55%

(b)   .02 = 2%

(c)   3.21 = 321%

(e)   .004 = .4%

Changing percents to decimals. To **change percents to decimals,**

1. Eliminate the percent sign.
2. Move the decimal point two places to the left (sometimes, adding zeros will be necessary).

Example 2: Change to decimals.

   (a)   24%   (b)   7%   (c)   250%

   (a) $24\% = .24$

   (b) $7\% = .07$

   (c) $250\% = 2.50 = 2.5$

Changing fractions to percents. To **change fractions to percents,**

   1.   Change to a decimal.
   2.   Change the decimal to a percent.

Example 3: Change to percents.

   (a) $\frac{1}{2}$ (b) $\frac{2}{5}$ (c) $\frac{5}{2}$ (d) $\frac{1}{20}$

   (a) $\frac{1}{2} = .5 = 50\%$

   (b) $\frac{2}{5} = .4 = 40\%$

   (c) $\frac{5}{2} = 2.5 = 250\%$

   (d) $\frac{1}{20} = .05 = 5\%$

Changing percents to fractions. To **change percents to fractions,**

   1.   Drop the percent sign.
   2.   Write over one hundred.
   3.   Reduce if necessary.

Example 4: Change to fractions.

(a) 13%   (b) 70%   (c) 45%   (d) 130%

(a)
$$13\% = \tfrac{13}{100}$$

(b)
$$70\% = \tfrac{70}{100} = \tfrac{7}{10}$$

(c)
$$45\% = \tfrac{45}{100} = \tfrac{9}{20}$$

(d)
$$130\% = \tfrac{130}{100} = \tfrac{13}{10} = 1\tfrac{3}{10}$$

Important equivalents that can save you time. Memorizing the following can eliminate computations.

$$\tfrac{1}{100} = .01 = 1\%$$

$$\tfrac{1}{10} = .1 = 10\%$$

$$\tfrac{1}{5} = \tfrac{2}{10} = .2 = .20 = 20\%$$

$$\tfrac{3}{10} = .3 = .30 = 30\%$$

$$\tfrac{2}{5} = \tfrac{4}{10} = .4 = .40 = 40\%$$

$$\tfrac{1}{2} = \tfrac{5}{10} = .5 = .50 = 50\%$$

$$\tfrac{3}{5} = \tfrac{6}{10} = .6 = .60 = 60\%$$

$$\tfrac{7}{10} = .7 = .70 = 70\%$$

$$\tfrac{4}{5} = \tfrac{8}{10} = .8 = .80 = 80\%$$

$$\tfrac{9}{10} = .9 = .90 = 90\%$$

$$\tfrac{1}{4} = \tfrac{25}{100} = .25 = 25\%$$

$$\tfrac{3}{4} = \tfrac{75}{100} = .75 = 75\%$$

$$\tfrac{1}{3} = .33\tfrac{1}{3} = 33\tfrac{1}{3}\%$$

$$\tfrac{2}{3} = .66\tfrac{2}{3} = 66\tfrac{2}{3}\%$$

$$\tfrac{1}{8} = .125 = .12\tfrac{1}{2} = 12\tfrac{1}{2}\%$$

$$\tfrac{3}{8} = .375 = .37\tfrac{1}{2} = 37\tfrac{1}{2}\%$$

$$\tfrac{5}{8} = .625 = .62\tfrac{1}{2} = 62\tfrac{1}{2}\%$$

$$\tfrac{7}{8} = .875 = .87\tfrac{1}{2} = 87\tfrac{1}{2}\%$$

$$\tfrac{1}{6} = .16\tfrac{2}{3} = 16\tfrac{2}{3}\%$$

$$\tfrac{5}{6} = .83\tfrac{1}{3} = 83\tfrac{1}{3}\%$$

$$1 = 1.00 = 100\%$$

$$2 = 2.00 = 200\%$$

$$3\tfrac{1}{2} = 3.5 = 3.50 = 350\%$$

Applications of Percents

Finding percent of a number. To **determine percent of a number,** change the percent to a fraction or decimal (whichever is easier for you) and multiply. Remember, the word *of* means multiply.

Example 5: Find the percents of these numbers.

(a) What is 20% of 80?   (c) What is $\frac{1}{2}$% of 18?

(b) What is 15% of 50?   (d) What is 70% of 20?

(a)   What is 20% of 80?

Using fractions,

$$20\% \text{ of } 80 = \tfrac{20}{100} \times \tfrac{80}{1} = \tfrac{1600}{100} = 16$$

Using decimals,

$$20\% \text{ of } 80 = .20 \times 80 = 16.00 = 16$$

(b)   What is 15% of 50?

Using fractions,

$$15\% \text{ of } 50 = \tfrac{15}{100} \times \tfrac{50}{1} = \tfrac{750}{100} = 7.5$$

Using decimals,

$$15\% \text{ of } 50 = .15 \times 50 = 7.5$$

(c)   What is $\frac{1}{2}$% of 18?

Using fractions,

$$\tfrac{1}{2}\% \text{ of } 18 = \frac{\frac{1}{2}}{100} \times 18 = \tfrac{1}{200} \times 18 = \tfrac{18}{200} = \tfrac{9}{100}$$

Note: $\dfrac{\frac{1}{2}}{100} = \tfrac{1}{2} \div \tfrac{100}{1} = \tfrac{1}{2} \times \tfrac{1}{100} = \tfrac{1}{200}$

Using decimals,

$$\tfrac{1}{2}\% \text{ of } 18 = .005 \times 18 = .09$$

(d) What is 70% of 20?

Using fractions,

$$70\% \text{ of } 20 = \tfrac{70}{100} \times \tfrac{20}{1} = \tfrac{1400}{100} = 14$$

Using decimals,

$$70\% \text{ of } 20 = .70 \times 20 = 14$$

Finding what percent one number is of another. One method to find **what percent one number is of another** is the **division method.** To use this method, simply take the number after the *of* and divide it into the number next to the *is.* Then change the answer to a percent.

Example 6: Find the percentages.

(a) 20 is what percent of 50?

(b) 27 is what percent of 90?

(a) 20 is what percent of 50?

$$\tfrac{20}{50} = \tfrac{2}{5} = .4 = 40\%$$

(b) 27 is what percent of 90?

$$\tfrac{27}{90} = \tfrac{3}{10} = .3 = 30\%$$

Another method to find what percent one number is of another is the **equation method.** Simply turn the question word-for-word into an equation. (To review solving simple equations, see page 145). For *what,* substitute the letter *x*; for *is,* substitute an *equal sign*

(=); for *of*, substitute a *multiplication sign* (×). Change percents to decimals or fractions, whichever find you find easier. Then solve the equation.

Example 7: Change each of the following into an equation and solve.

  (a)   10 is what percent of 50?

  (b)   15 is what percent of 60?

  (a)   10 is what percent of 50?

$$10 = x(50)$$

$$\frac{10}{50} = \frac{x(50)}{50}$$

$$\frac{10}{50} = \frac{x(\cancel{50})}{\cancel{50}}$$

$$\tfrac{10}{50} = x$$

$$\tfrac{1}{5} = x$$

$$20\% = x$$

  (b)   15 is what percent of 60?

$$15 = x(60)$$

$$\frac{15}{60} = \frac{x(60)}{60}$$

$$\frac{15}{60} = \frac{x(\cancel{60})}{\cancel{60}}$$

$$\tfrac{15}{60} = x$$

$$\tfrac{1}{4} = x$$

$$25\% = x$$

Finding a number when a percent of it is known. One method used to **find a number when a percent of it is known** is the **division method.** To use this method, simply take the number of percent, change it into a decimal, and divide that into the other number.

Example 8: Find the number.

   (a)   15 is 50% of what number?

   (b)   20 is 40% of what number?

(a)   15 is 50% of what number?

$$\frac{15}{.50} = 30 \quad \text{or} \quad 15 \div \tfrac{1}{2} = \tfrac{15}{1} \times \tfrac{2}{1} = 30$$

(b)   20 is 40% of what number?

$$\frac{20}{.40} = 50 \quad \text{or} \quad 20 \div \tfrac{2}{5} = \tfrac{20}{1} \times \tfrac{5}{2} = \tfrac{100}{2} = 50$$

     Another method to find a number when a percent of it is known is the **equation method.** Simply turn the question word-for-word into an equation. For *what*, substitute the letter x; for *is*, substitute an *equal sign* (=); for *of*, substitute a *multiplication sign* (×). Change percents to decimals or fractions, whichever you find easier. Then solve the equation.

Example 9: Find the number.

(a) 30 is 20% of what number?

(b) 40 is 80% of what number?

(a) 　　　　　　30 is 20% of what number?

$$30 = .20x \qquad \text{or} \qquad 30 = (\tfrac{1}{5})x$$

$$\frac{30}{.20} = \frac{.20x}{.20} \qquad\qquad (\tfrac{5}{1})(\tfrac{30}{1}) = (\tfrac{1}{5})x(\tfrac{5}{1})$$

$$\frac{30}{.20} = \frac{.\cancel{20}x}{\cancel{.20}} \qquad\qquad 150 = x$$

$$\frac{30}{.20} = x$$

$$150 = x$$

(b) 　　　　　　40 is 80% of what number?

$$40 = .80x \qquad \text{or} \qquad 40 = (\tfrac{4}{5})x$$

$$\frac{.40}{80} = \frac{.80x}{.80} \qquad\qquad (\tfrac{5}{4})(\tfrac{40}{1}) = (\tfrac{4}{5})x(\tfrac{5}{4})$$

$$\frac{.40}{.80} = \frac{.\cancel{80}x}{\cancel{80}} \qquad\qquad 50 = x$$

$$\frac{40}{.80} = x$$

$$50 = x$$

Percent—proportion method. Another simple method commonly used to solve any of the three types of percent problems is the **proportion**, or **is/of, method.** First set up a blank proportion and then fill in the empty spaces by using the following steps.

$$\frac{?}{?} = \frac{?}{?}$$

1. Whatever is next to the percent (%) is put over 100. (The word *what* is the unknown, or *x*.)

2. Whatever comes immediately after the word *of* goes on the bottom of one side of the proportion.

3. Whatever is left (comes next to the word *is*) goes on top, on one side of the proportion.

4. Then solve the problem.

Example 10: 30 is what percent of 50?

Set up a blank proportion.

$$\frac{?}{?} = \frac{?}{?}$$

30 is what percent of 50?

Step 1. $\qquad\qquad \dfrac{x}{100} = \dfrac{?}{?}$

Step 2: $\qquad\qquad \dfrac{x}{100} = \dfrac{?}{50}$

Step 3: $\qquad\qquad \dfrac{x}{100} = \dfrac{30}{50}$

Step 4: $\qquad\qquad \dfrac{x}{100} = \dfrac{30}{50} = \dfrac{60}{100} = 60\%$

In this particular problem, however, it can be observed quickly that $\frac{30}{50} = \frac{60}{100} = 60\%$, so solving mechanically as shown would not be time effective.

The proportion method works for the three basic types of percent questions.

1. 30 is what percent of 50?

2. 60 is 20% of what number?

3. What number is 15% of 30? (In this type it is probably easier to simply multiply the numbers.)

Example 11: Solve using the proportion method.

(a) 40 is what percent of 200?

(b) What is percent of 25 is 10?

(c) 60 is 20% of what number?

(d) What number is 15% of 30?

(a)   40 is what percent of 200?

$$\frac{x}{100} = \frac{40}{200}$$

Since $\frac{40}{200}$ can be reduced to $\frac{20}{100}$, $x = 20$. So 40 is 20 percent of 200. (This particular problem does not need to be solved mechanically.)

(b)   What percent of 25 is 10?

$$\frac{x}{100} = \frac{10}{25}$$

$$25x = 1000$$

$$\frac{\overset{1}{\cancel{25}}x}{\cancel{25}} = \frac{\overset{40}{\cancel{1000}}}{\cancel{25}}$$

$$x = 40$$

So 40 percent of 25 is 10. (You could also solve this by observing that $4 \times 25 = 100$; therefore, $\frac{40}{100} = \frac{10}{25}$.)

(c)   60 is 20% of what number?

$$\frac{20}{100} = \frac{60}{x} \qquad \text{or} \qquad \frac{20}{100} = \frac{60}{x}$$

$$20x = 6000 \qquad\qquad \frac{1}{5} = \frac{60}{x}$$

$$\frac{\cancel{20}x}{\cancel{20}} = \frac{6000}{20} \qquad\qquad x = 300$$

$$x = 300$$

So 60 is 20% of 300.

(d)   What number is 15% of 30?

$$\frac{15}{100} = \frac{x}{30}$$

$$450 = 100x$$

$$\frac{450}{100} = \frac{\cancel{100}x}{\cancel{100}}$$

$$x = 4.5$$

So 4.5 is 15% of 30.

Finding percent increase or percent decrease. To find percent change (increase or decrease), use this formula.

$$\frac{\text{change}}{\text{starting point}} = \text{percent change}$$

Example 12: Find the percent change.

(a)   What is the percent decrease of a $500 item on sale for $400?

(b)   What is the percent increase of Jon's salary if it went from $150 a month to $200 a month?

(c)   What is the percent change from 2100 to 1890?

(a)   percent decrease from 500 to 400

$$\text{change} = 500 - 400 = 100$$

$$\frac{\text{change}}{\text{starting point}} = \frac{100}{500} = \frac{1}{5} = 20\% \text{ decrease}$$

(b)   percent increase from 150 to 200

$$\text{change} = 200 - 150 = 50$$

$$\frac{\text{change}}{\text{starting point}} = \frac{50}{150} = \frac{1}{3} = 33\tfrac{1}{3}\% \text{ increase}$$

(c)   percent change from 2100 to 1890

$$\text{change} = 2100 - 1890 = 210$$

$$\frac{\text{change}}{\text{starting point}} = \frac{210}{2100} = \frac{1}{10} = 10\% \text{ change}$$

Note that the terms *percentage rise, percentage difference,* and *percentage change* are the same as *percent change.*

Integers

The term **integers** refers to positive whole numbers, negative whole numbers, and zero—no fractions or decimals. . . . $-3, -2, -1, 0, 1, 2, 3, \ldots$ are integers. They are signed whole numbers.

Number lines. On a **number line,** numbers to the right of 0 are positive. Numbers to the left of 0 are negative, as shown in Figure 1.

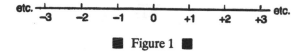

etc. $-3 \quad -2 \quad -1 \quad 0 \quad +1 \quad +2 \quad +3$ etc.

■ Figure 1 ■

This figure shows only the integers on the number line.

Given any two numbers on a number line, the one on the right is always larger, regardless of its sign (positive or negative).

Addition of integers. When **adding two integers with the same sign,** (either both positive or both negative), add the numbers and keep the sign.

Example 1: Add the following.

(a) $\begin{array}{r} +3 \\ + +5 \\ \hline \end{array}$ (c) $\begin{array}{r} +4 \\ + +12 \\ \hline \end{array}$

(b) $\begin{array}{r} -6 \\ + -3 \\ \hline \end{array}$ (d) $\begin{array}{r} -8 \\ + -9 \\ \hline \end{array}$

(a)
$$\begin{array}{r} +3 \\ ++5 \\ \hline +8 \end{array}$$

(b)
$$\begin{array}{r} -6 \\ +-3 \\ \hline -9 \end{array}$$

(c)
$$\begin{array}{r} +4 \\ ++12 \\ \hline +16 \end{array}$$

(d)
$$\begin{array}{r} -8 \\ +-9 \\ \hline -17 \end{array}$$

When **adding integers with different signs,** (one positive and one negative), subtract the numbers and keep the sign from the "larger" one (that is, the number that is larger if you disregard the positive or negative sign).

Example 2: Add the following.

(a) $\begin{array}{r} +8 \\ +-9 \\ \hline \end{array}$ (b) $\begin{array}{r} -30 \\ ++45 \\ \hline \end{array}$

(a)
$$\begin{array}{r} +8 \\ +-9 \\ \hline -1 \end{array}$$

(b)
$$\begin{array}{r} -30 \\ ++45 \\ \hline +15 \end{array}$$

Integers may also be added "horizontally."

Example 3: Add the following.

(a) +8 + 11   (b) −15 + 7   (c) 5 + (−3)   (d) −21 + 6

(a)   +8 + 11 = +19

(b)   −15 + 7 = −8

(c)   5 + (−3) = +2

(d)   −21 + 6 = −15

Subtraction of integers. To **subtract positive and/or negative integers,** just change the sign of the number being subtracted and then use the rules for adding integers.

Example 4: Subtract the following.

(a)   +12   (c)   −19
     − +4   − +6

(b)   −14   (d)   +20
     − −4   − −3

(a)
$$\begin{array}{r} +12 = \ +12 \\ -+4 = \ +-4 \\ \hline +8 \end{array}$$

(b)
$$\begin{array}{r} -14 = \ -14 \\ --4 = \ ++4 \\ \hline -10 \end{array}$$

(c)
$$-19 = -19$$
$$\underline{-+6 = +-6}$$
$$-25$$

(d)
$$+20 = +20$$
$$\underline{--3 = ++3}$$
$$+23$$

Subtracting positive and/or negative integers may also be done "horizontally."

Example 5: Subtract the following.

(a) $+12 - (+4)$ (c) $-20 - (+3)$

(b) $+16 - (-6)$ (d) $-5 - (-2)$

(a) $+12 - (+4) = +12 + (-4) = 8$

(b) $+16 - (-6) = +16 + (+6) = 22$

(c) $-20 - (+3) = -20 + (-3) = -23$

(d) $-5 - (-2) = -5 + (+2) = -3$

Minus preceding parenthesis. If a **minus precedes a parenthesis,** it means that everything within the parentheses is to be subtracted. Therefore, using the same rule as in subtraction of integers, simply change every sign within the parentheses to its opposite and then add.

Example 6: Subtract the following.

   (a) $9 - (+3 - 5 + 7 - 6)$ (b) $20 - (+35 - 50 + 100)$

   (a) $9 - (+3 - 5 + 7 - 6) = 9 + (-3 + 5 - 7 + 6)$

$$= 9 + (+1)$$
$$= 10$$

   (b) $20 - (+35 - 50 + 100) = 20 + (-35 + 50 - 100)$

$$= 20 + (-85)$$
$$= -65$$

Or you could total the numbers within the parentheses by first adding the positive numbers together, then adding the negative numbers together, then combining, and finally subtracting.

Example 7: Subtract the following.

   (a) $9 - (+3 - 5 + 7 - 6)$
   (b) $20 - (+35 - 50 + 100)$
   (c) $3 - (1 - 4)$

   (a) $9 - (+3 - 5 + 7 - 6) = 9 - (+10 - 11)$

$$= 9 - (-1)$$
$$= 9 + (+1)$$
$$= 10$$

   (b) $20 - (+35 - 50 + 100) = 20 - (135 - 50)$

$$= 20 - (85)$$
$$= -65$$

(c)   Remember, if there is no sign given, the number is
understood to be positive.

$$3 - (1 - 4) = 3 - (+1 - 4)$$
$$= 3 + (-1 + 4)$$
$$= 3 + 3$$
$$= 6$$

Or $$3 - (1 - 4) = 3 - (-3)$$
$$= 3 + 3$$
$$= 6$$

Multiplying and dividing integers. To **multiply or divide integers,**
treat them just like regular numbers but remember this rule: An odd
number of negative signs will produce a negative answer. An even
number of negative signs will produce a positive answer.

Example 8: Multiply or divide the following.

(a) $(-4) \times (-7)$ (d) $\dfrac{-16}{-4}$

(b) $(-2) \times (+3) \times (-8) \times (-1)$ (e) $\dfrac{-48}{+3}$

(c) $(+5) \times (-6) \times (-2) \times (+4) \times (+3)$ (f) $\dfrac{+28}{-7}$

(a) $(-4) \times (-7) = +28$

(b) $(-2) \times (+3) \times (-8) \times (-1) = -48$

(c) $(+5) \times (-6) \times (-2) \times (+4) \times (+3) = +720$

(d)
$$\frac{-16}{-4} = +4$$

(e)
$$\frac{-48}{+3} = -16$$

(f)
$$\frac{+28}{-7} = -4$$

Absolute value. The numerical value when direction or sign is not considered is called the **absolute value.** The absolute value of a number is written $|3| = 3$ and $|-4| = 4$. The absolute value of a number is always positive except when the number is 0.

Example 9: Give the value.

(a) $|5|$ (b) $|-8|$ (c) $|3 - 9|$ (d) $3 - |-6|$

(a) $$|5| = 5$$

(b) $$|-8| = 8$$

(c) $$|3 - 9| = |-6| = 6$$

(d) $$3 - |-6| = 3 - 6 = -3$$

Note that absolute value is taken first.

Rationals (Signed Numbers Including Fractions)

As mentioned earlier, integers are positive and negative whole numbers and zero. When the fractions and decimals between the integers are included, the complete group of numbers is referred to as **rational numbers.** They are signed numbers including fractions. A more technical definition of a rational number is any number that can be written as a fraction with the numerator being a whole number and the denominator being a natural number. Notice that fractions can be placed on the number line as shown in Figure 2.

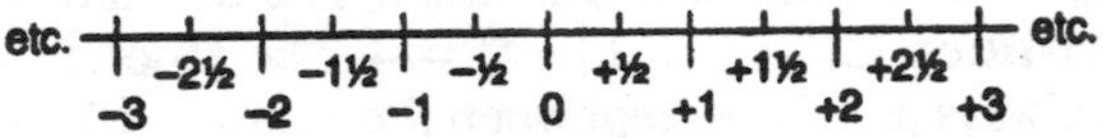

etc. ┼──┼──┼──┼──┼──┼──┼──┼──┼──┼──┼──┼──┼ etc.

$$-3 \quad -2\tfrac{1}{2} \quad -2 \quad -1\tfrac{1}{2} \quad -1 \quad -\tfrac{1}{2} \quad 0 \quad +\tfrac{1}{2} \quad +1 \quad +1\tfrac{1}{2} \quad +2 \quad +2\tfrac{1}{2} \quad +3$$

■ Figure 2 ■

Negative fractions. Fractions may be negative as well as positive, as you can see in Figure 2. Since fractions (positive fractions) were discussed in an earlier section, the focus here will be on negative fractions. Negative fractions are typically written

$$-\tfrac{3}{4} \quad \text{not} \quad \tfrac{-3}{4} \quad \text{or} \quad \tfrac{3}{-4}$$

although they are all equal. $\quad -\tfrac{3}{4} = \tfrac{-3}{4} = \tfrac{3}{-4}$

Adding positive and negative fractions. The rules for signs when adding integers applies to fractions as well. Remember, to add fractions, you must first get a common denominator.

Example 10: Add the following.

(a) $-\tfrac{1}{5} + \tfrac{3}{5}$ (b) $-\tfrac{2}{3} + (-\tfrac{1}{4})$ (c) $-\tfrac{1}{2} + \tfrac{3}{8}$

(a) $$-\tfrac{1}{5} + \tfrac{3}{5} = \tfrac{2}{5}$$

(b)

$$-\tfrac{2}{3} = \quad -\tfrac{8}{12}$$
$$+-\tfrac{1}{4} = +-\tfrac{3}{12}$$
$$\overline{\qquad\qquad -\tfrac{11}{12}}$$

(c)

$$-\tfrac{1}{2} + \tfrac{3}{8} = -\tfrac{4}{8} + \tfrac{3}{8} = -\tfrac{1}{8}$$

Adding positive and negative mixed numbers. The rules for signs when adding integers applied to mixed numbers as well.

Example 11: Add the following.

(a) $-2\tfrac{3}{4} + (-5\tfrac{1}{5})$ (b) $-4\tfrac{1}{2} + 2\tfrac{1}{3}$

(a)

$$-2\tfrac{3}{4} = \quad -2\tfrac{15}{20}$$
$$+-5\tfrac{1}{5} = +-5\tfrac{4}{20}$$
$$\overline{\qquad\qquad -7\tfrac{19}{20}}$$

(b)

$$-4\tfrac{1}{2} + 2\tfrac{1}{3} = -4\tfrac{3}{6} + 2\tfrac{2}{6} = -2\tfrac{1}{6}$$

Subtracting positive and negative fractions. The rules for signs when subtracting integers applies to fractions as well. Remember, to subtract fractions, you must first get a common denominator.

Example 12: Subtract the following.

(a) $+\tfrac{7}{10} - (-\tfrac{1}{5})$ (b) $+\tfrac{2}{7} - (-\tfrac{1}{4})$ (c) $+\tfrac{1}{6} - \tfrac{3}{4}$

(a)

$$+\tfrac{7}{10} = \quad +\tfrac{7}{10} = +\tfrac{7}{10}$$
$$--\tfrac{1}{5} = ++\tfrac{1}{5} = +\tfrac{2}{10}$$
$$\overline{\qquad\qquad\qquad\qquad +\tfrac{9}{10}}$$

(b) $\quad +\frac{2}{7} - (-\frac{1}{4}) = +\frac{8}{28} - (-\frac{7}{28}) = +\frac{8}{28} + \frac{7}{28} = +\frac{15}{28}$

or

$$+\frac{2}{7} - (-\frac{1}{4}) = +\frac{2}{7} + \frac{1}{4} = +\frac{8}{28} + \frac{7}{28} = +\frac{15}{28}$$

(c) $\quad +\frac{1}{6} - \frac{3}{4} = +\frac{2}{12} - \frac{9}{12} = +\frac{2}{12} + (-\frac{9}{12}) = -\frac{7}{12}$

Subtracting positive and negative mixed numbers. The rules for signs when subtracting integers applies to mixed numbers as well. Remember, to subtract mixed numbers, you must first get a common denominator. If borrowing from a column is necessary, be cautious of simple mistakes.

Example 13: Subtract the following.

(a) $+2\frac{1}{4} - (+3\frac{1}{2})$ (b) $-5\frac{1}{8} - (-2\frac{1}{3})$ (c) $+4\frac{2}{7} - 7\frac{2}{5}$

(a)
$$\begin{array}{r} +2\frac{1}{4} = +2\frac{1}{4} \\ -+3\frac{1}{2} = +-3\frac{2}{4} \\ \hline -1\frac{1}{4} \end{array} \quad \text{same as} \quad \begin{array}{r} -3\frac{2}{4} \\ +2\frac{1}{4} \\ \hline -1\frac{1}{4} \end{array}$$

(b)
$$\begin{array}{r} -5\frac{1}{8} = -5\frac{1}{8} = -5\frac{3}{24} = -4\frac{27}{24} \\ --2\frac{1}{3} = +2\frac{1}{3} = +2\frac{8}{24} = +2\frac{8}{24} \\ \hline -2\frac{19}{24} \end{array}$$

(c) $\quad +4\frac{2}{7} - 7\frac{2}{5} = +4\frac{10}{35} - 7\frac{14}{35} = -3\frac{4}{35}$

Problems such as those above are usually most easily done by stacking the number with the "largest" value (disregarding the positive or negative sign) on top, subtracting, and keeping the sign of the number with the "largest" value.

Multiplying positive and negative fractions. The rules for signs when multiplying integers applies to fractions as well. Remember, to multiply fractions, multiply the numerators and then multiply the denominators. Always reduce to lowest terms if possible.

Example 14: Multiply the following.

(a) $(-\frac{2}{3}) \times (+\frac{1}{5})$ (b) $(-\frac{5}{6}) \times (-\frac{7}{9})$ (c) $(+\frac{3}{8}) \times (-\frac{2}{7})$

(a) $(-\frac{2}{3}) \times (+\frac{1}{5}) = -\frac{2}{15}$

(b) $(-\frac{5}{6}) \times (-\frac{7}{9}) = +\frac{35}{54}$

(c) $(+\frac{3}{8}) \times (-\frac{2}{7}) = -\frac{6}{56} = -\frac{3}{28}$

Canceling. You can **cancel** when multiplying positive and negative fractions. Simply cancel as you did when multiplying positive fractions, but pay special attention to the signs involved. Follow the rules for signs when multiplying integers to obtain the proper sign. Remember, no sign means that a positive sign is understood.

Example 15: Multiply the following.

(a) $(-\frac{2}{3}) \times (\frac{5}{12})$ (b) $(-\frac{5}{6}) \times (-\frac{3}{10})$ (c) $(\frac{1}{5}) \times (-\frac{15}{16})$

(a) $(-\frac{2}{3}) \times (\frac{5}{12}) = \left(-\frac{\overset{1}{\cancel{2}}}{3}\right) \times \left(\frac{5}{\underset{6}{\cancel{12}}}\right)$

$= -\frac{5}{18}$

(b) $(-\frac{5}{6}) \times (-\frac{3}{10}) = \left(-\frac{\cancel{5}^{1}}{\cancel{6}_{2}}\right) \times \left(-\frac{\cancel{3}^{1}}{\cancel{10}_{2}}\right)$

$$= +\frac{1}{4}$$

(c) $(\frac{1}{5}) \times (-\frac{15}{16}) = \left(\frac{1}{\cancel{5}_{1}}\right) \times \left(-\frac{\cancel{15}^{3}}{16}\right)$

$$= -\frac{3}{16}$$

Multiplying positive and negative mixed numbers. Follow the rules for signs when multiplying integers to get the proper sign. Remember, before multiplying mixed numbers, you must first change them to improper fractions.

Example 16: Multiply the following.

(a) $(-3\frac{1}{4}) \times (2\frac{1}{2})$ (b) $(6\frac{1}{2}) \times (-1\frac{1}{6})$ (c) $(-5\frac{1}{4}) \times (-1\frac{1}{3})$

(a) $(-3\frac{1}{4}) \times (2\frac{1}{2}) = (-\frac{13}{4}) \times (\frac{5}{2})$

$$= -\frac{65}{8}$$

$$= -8\frac{1}{8}$$

(b) $(6\frac{1}{2}) \times (-1\frac{1}{6}) = (\frac{13}{2}) \times (-\frac{7}{6})$

$$= -\frac{91}{12}$$

$$= -7\frac{7}{12}$$

(c)
$$(-5\tfrac{1}{4}) \times (-1\tfrac{1}{3}) = (-\tfrac{21}{4}) \times (-\tfrac{4}{3})$$

$$= \left(-\frac{\overset{7}{\cancel{21}}}{\cancel{4}}\right) \times \left(-\frac{\cancel{4}}{\underset{1}{\cancel{3}}}\right)$$

$$= \tfrac{7}{1}$$

$$= 7$$

Dividing positive and negative fractions. Follow the rules for signs when dividing integers to get the proper sign. Remember, when dividing fractions, first invert the divisor and then multiply.

Example 17: Divide the following.

(a) $(-\tfrac{2}{3}) \div (\tfrac{1}{4})$ (b) $(-\tfrac{2}{5}) \div (-\tfrac{3}{4})$ (c) $(\tfrac{1}{6}) \div (-\tfrac{2}{3})$

(a)
$$(-\tfrac{2}{3}) \div (\tfrac{1}{4}) = (-\tfrac{2}{3}) \times (\tfrac{4}{1})$$

$$= -\tfrac{8}{3}$$

$$= -2\tfrac{2}{3}$$

(b)
$$(-\tfrac{2}{5}) \div (-\tfrac{3}{4}) = (-\tfrac{2}{5}) \times (-\tfrac{4}{3})$$

$$= \tfrac{8}{15}$$

(c)
$$(\tfrac{1}{6}) \div (-\tfrac{2}{3}) = (\tfrac{1}{6}) \times (-\tfrac{3}{2})$$

$$= \left(\frac{1}{\underset{2}{\cancel{6}}}\right) \times \left(-\frac{\overset{1}{\cancel{3}}}{2}\right)$$

$$= -\tfrac{1}{4}$$

Dividing positive and negative mixed numbers. Follow the rules for signs when dividing integers to get the proper sign. Remember, before dividing mixed numbers, you must first change them to improper fractions. Then you must invert the divisor and multiply.

Example 18: Divide the following.

(a) $(-2\frac{1}{2}) \div (\frac{1}{3})$ (b) $(-3\frac{1}{2}) \div (-4\frac{1}{3})$ (c) $(5\frac{1}{4}) \div (-3\frac{3}{5})$

(a)
$$(-2\tfrac{1}{2}) \div (\tfrac{1}{3}) = (-\tfrac{5}{2}) \div (\tfrac{1}{3})$$
$$= (-\tfrac{5}{2}) \times (\tfrac{3}{1})$$
$$= -\tfrac{15}{2}$$
$$= -7\tfrac{1}{2}$$

(b)
$$(-3\tfrac{1}{2}) \div (-4\tfrac{1}{3}) = (-\tfrac{7}{2}) \div (-\tfrac{13}{3})$$
$$= (-\tfrac{7}{2}) \times (-\tfrac{3}{13})$$
$$= \tfrac{21}{26}$$

(c)
$$(5\tfrac{1}{4}) \div (-3\tfrac{3}{5}) = (\tfrac{21}{4}) \div (-\tfrac{18}{5})$$
$$= (\tfrac{21}{4}) \times (-\tfrac{5}{18})$$
$$= \left(\frac{\overset{7}{\cancel{21}}}{4}\right) \times \left(-\frac{5}{\underset{6}{\cancel{18}}}\right)$$
$$= -\tfrac{35}{24}$$
$$= -1\tfrac{11}{24}$$

Powers and Exponents

Exponents. An **exponent** is a positive or negative number placed above and to the right of a quantity. It expresses the **power** to which the quantity is to be raised or lowered. In 4^3, 3 is the exponent. It shows that 4 is to be used as a factor three times: $4 \times 4 \times 4$ (multiplied by itself twice). 4^3 is read as *four to the third power* (or *four cubed* as discussed below).

$$2^4 = 2 \times 2 \times 2 \times 2 = 16$$
$$3^2 = 3 \times 3 = 9$$
$$5^3 = 5 \times 5 \times 5 = 125$$

Remember that $x^1 = x$ and $x^0 = 1$ when x is any number (other than 0).

$$2^1 = 2 \qquad 2^0 = 1$$
$$3^1 = 3 \qquad 3^0 = 1$$
$$4^1 = 4 \qquad 4^0 = 1$$

Negative exponents. If the **exponent is negative,** such as 4^{-2}, then the number and exponent may be dropped under the number 1 in a fraction to remove the negative sign.

Example 1: Simplify the following by removing the exponents.

(a) 4^{-2} (b) 5^{-3} (c) 2^{-4} (d) 3^{-1}

(a) $$4^{-2} = \frac{1}{4^2} = \frac{1}{16}$$

(b) $$5^{-3} = \frac{1}{5^3} = \frac{1}{125}$$

(c) $$2^{-4} = \frac{1}{2^4} = \frac{1}{16}$$

(d) $$3^{-1} = \frac{1}{3^1} = \frac{1}{3}$$

Squares and cubes. Two specific types of powers should be noted, **squares** and **cubes**. To **square** a number, just multiply it by itself (the exponent would be 2). For example 6 squared (written 6^2) is 6×6, or 36. 36 is called a **perfect square** (the square of a whole number). Following is a partial list of perfect squares.

$$0^2 = 0 \qquad 5^2 = 25 \qquad 9^2 = 81$$
$$1^2 = 1 \qquad 6^2 = 36 \qquad 10^2 = 100$$
$$2^2 = 4 \qquad 7^2 = 49 \qquad 11^2 = 121$$
$$3^2 = 9 \qquad 8^2 = 64 \qquad 12^2 = 144$$
$$4^2 = 16$$

To **cube** a number, just multiply it by itself twice (the exponent would be 3). For example, 5 cubed (written 5^3) is $5 \times 5 \times 5$, or 125. 125 is called a **perfect cube** (the cube of a whole number). Following is a partial list of perfect cubes.

$$0^3 = 0 \qquad\qquad 4^3 = 4 \times 4 \times 4 = 64$$

$$1^3 = 1 \times 1 \times 1 = 1 \qquad 5^3 = 5 \times 5 \times 5 = 125$$

$$2^3 = 2 \times 2 \times 2 = 8 \qquad 6^3 = 6 \times 6 \times 6 = 216$$

$$3^3 = 3 \times 3 \times 3 = 27 \qquad 7^3 = 7 \times 7 \times 7 = 343$$

Operations with powers and exponents. To **multiply** two numbers with exponents, **if the base numbers are the same,** simply keep the base number and add the exponents.

Example 2: Multiply the following, leaving the answers with exponents.

(a) $2^3 \times 2^5$ (b) $3^2 \times 3^5$ (c) $5^4 \times 5^7$

(a)
$$2^3 \times 2^5 = 2^{(3+5)} = 2^8$$
$$(2 \times 2 \times 2) \times (2 \times 2 \times 2 \times 2 \times 2) = 2^8$$

(b)
$$3^2 \times 3^5 = 3^{(2+5)} = 3^7$$

(c)
$$5^4 \times 5^7 = 5^{(4+7)} = 5^{11}$$

To **divide** two numbers with exponents, **if the base numbers are the same,** simply keep the base number and subtract the second exponent from the first, or the exponent of the denominator from the exponent of the numerator.

Example 3: Divide the following, leaving the answers with exponents.

(a) $5^6 \div 5^2$ (b) $\dfrac{8^7}{8^3}$

(a) $$5^6 \div 5^2 = 5^{(6-2)} = 5^4$$

(b) $$\dfrac{8^7}{8^3} = 8^{(7-3)} = 8^4$$

To **multiply** or **divide** numbers with exponents, **if the base numbers are different,** you must simplify each number with an exponent first and then perform the operation.

Example 4: Simplify and perform the operation indicated.

(a) $2^3 \times 3^2$ (b) $6^2 \div 2^3$

(a) $$2^3 \times 3^2 = 8 \times 9 = 72$$

(b) $$6^2 \div 2^3 = 36 \div 8 = 4\tfrac{4}{8} = 4\tfrac{1}{2}$$

For problems such as those in Example 4, some shortcuts are possible.

To **add** or **subtract** numbers with exponents, **whether the base numbers are the same or different,** you must simplify each number with an exponent first and then perform the indicated operation.

Example 5: Simplify and perform the operation indicated.

  (a) $3^2 - 2^3$ (b) $5^2 + 3^3$ (c) $4^2 + 9^3$ (d) $2^3 - 2^2$

(a) $$3^2 - 2^3 = 9 - 8 = 1$$

(b) $$5^2 + 3^3 = 25 + 27 = 52$$

(c) $$4^2 + 9^3 = 16 + 729 = 745$$

(d) $$2^3 - 2^2 = 8 - 4 = 4$$

If a **number with an exponent is taken to another power** $(4^2)^3$, simply keep the original base number and multiply the exponents.

Example 6: Multiply the following and leave the answers with exponents.

  (a) $(6^3)^2$ (b) $(3^2)^4$ (c) $(5^4)^3$

(a) $$(6^3)^2 = 6^{(3 \times 2)} = 6^6$$

(b) $$(3^2)^4 = 3^{(2 \times 4)} = 3^8$$

(c) $$(5^4)^3 = 5^{(4 \times 3)} = 5^{12}$$

Square Roots and Cube Roots

Note that **square** and **cube roots** and operations with them are often included in algebra books.

Square roots. To find the **square root** of a number, you want to find some number that when multiplied by itself gives you the original number. In other words, to find the square root of 25, you want to find the number that when multiplied by itself gives you 25. The square root of 25, then, is 5. The symbol for square root is $\sqrt{}$. Following is a partial list of perfect (whole number) square roots.

$$\sqrt{0} = 0 \qquad \sqrt{16} = 4 \qquad \sqrt{64} = 8$$
$$\sqrt{1} = 1 \qquad \sqrt{25} = 5 \qquad \sqrt{81} = 9$$
$$\sqrt{4} = 2 \qquad \sqrt{36} = 6 \qquad \sqrt{100} = 10$$
$$\sqrt{9} = 3 \qquad \sqrt{49} = 7$$

Special note: If no sign (or a positive sign) is placed in front of the square root, then the positive answer is required. Only if a negative sign is in front of the square root is the negative answer required. This notation is used in many texts and will be adhered to in this book. Therefore,

$$\sqrt{4} = 2 \quad \text{and} \quad -\sqrt{4} = -2$$

Cube roots. To find the **cube root** of a number, you want to find some number that when multiplied by itself twice gives you the original number. In other words, to find the cube root of 8, you want to find the number that when multiplied by itself twice gives you 8. The cube root of 8, then, is 2, since $2 \times 2 \times 2 = 8$. Notice that the symbol for cube root is the radical sign with a small three (called the **index**) above and to the left $\sqrt[3]{}$. Other roots are similarly defined and identified by the index given. (In square root, an index of two is understood and usually not written.) Following is a partial list of **perfect** (whole number) **cube roots.**

$$\sqrt[3]{0} = 0 \qquad \sqrt[3]{27} = 3$$
$$\sqrt[3]{1} = 1 \qquad \sqrt[3]{64} = 4$$
$$\sqrt[3]{8} = 2 \qquad \sqrt[3]{125} = 5$$

Approximating square roots. To find the square root of a number that is not a perfect square, it will be necessary to find an *approximate* answer by using the procedure given in Example 7.

Example 7: Approximate $\sqrt{42}$.

$\sqrt{42}$ is between $\sqrt{36}$ and $\sqrt{49}$.

$$\sqrt{36} < \sqrt{42} < \sqrt{49}$$
$$\sqrt{36} = 6$$
$$\sqrt{49} = 7$$

Therefore, $\qquad\qquad 6 < \sqrt{42} < 7$

Since 42 is almost halfway between 36 and 49, $\sqrt{42}$ is almost halfway between $\sqrt{36}$ and $\sqrt{49}$. So $\sqrt{42}$ is approximately 6.5. To check, multiply.

$$6.5 \times 6.5 = 42.25$$

or about 42.

Example 8: Approximate $\sqrt{71}$.

$$\sqrt{64} < \sqrt{71} < \sqrt{81}$$
$$8 < \sqrt{71} < 9$$

Since $\sqrt{71}$ is slightly closer to $\sqrt{64}$ than it is to $\sqrt{81}$,

$$8 < 8.4 < 9$$
$$\sqrt{71} \approx 8.4$$

Check the answer.

$$
\begin{array}{r}
8.4 \\
\times\, 8.4 \\
\hline
336 \\
672 \\
\hline
70.56 \approx 71
\end{array}
$$

Example 9: Approximate $\sqrt{\frac{300}{15}}$.

First, perform the operation under the radical.

$$\sqrt{\tfrac{300}{15}} = \sqrt{20}$$

$$\sqrt{16} < \sqrt{20} < \sqrt{25}$$

$$4 < \sqrt{20} < 5$$

Since $\sqrt{20}$ is slightly closer to $\sqrt{16}$ than it is to $\sqrt{25}$,

$$4 < 4.4 < 5$$

$$\sqrt{\tfrac{300}{15}} \approx 4.4$$

Square roots of nonperfect squares can be approximated, looked up in tables, or found by using a calculator. You may wish to keep these two in mind, since they are commonly used.

$$\sqrt{2} \approx 1.414 \qquad \sqrt{3} \approx 1.732$$

Simplifying square roots. Sometimes, you will have to **simplify** square roots, or write them in simplest form. In fractions, $\frac{2}{4}$ can be reduced to $\frac{1}{2}$. In square roots, $\sqrt{32}$ can be simplified to $4\sqrt{2}$. To **simplify a square root,** first factor the number under the $\sqrt{}$ into two factors, one of which is the largest possible perfect square. (Perfect square numbers are 1, 4, 9, 16, 25, 49, . . .)

Example 10: Simplify $\sqrt{32}$.

$$\sqrt{32} = \sqrt{16 \times 2}$$

Then, take the square root of the perfect square number.

$$\sqrt{16 \times 2} = \sqrt{16} \times \sqrt{2} = 4 \times \sqrt{2}$$

and, finally, write as a single expression

$$4\sqrt{2}$$

Example 11: Simplify $\sqrt{24}$.

$$\sqrt{24} = \sqrt{4 \times 6}$$
$$= \sqrt{4} \times \sqrt{6}$$
$$= 2 \times \sqrt{6}$$
$$= 2\sqrt{6}$$

To check, square the number on the outside of the radical and multiply it by the number on the inside.

$$2^2 \times 6 = 4 \times 6$$
$$= 24 \quad \checkmark$$

Example 12: Simplify $\sqrt{75}$

$$\sqrt{75} = \sqrt{25 \times 3}$$
$$= \sqrt{25} \times \sqrt{3}$$
$$= 5 \times \sqrt{3}$$
$$= 5\sqrt{3}$$

Remember that most square roots cannot be simplified, as they are already in simplest form, such as $\sqrt{7}$, $\sqrt{10}$, and $\sqrt{15}$

Powers of Ten

Since our number system is based on **powers of ten,** as mentioned earlier, you should understand the notation and how to work with these powers

$$10^0 = 1$$
$$10^1 = 10$$
$$10^2 = 10 \times 10 = 100$$
$$10^3 = 10 \times 10 \times 10 = 1000$$

and so on

$$10^{-1} = \tfrac{1}{10} = .1$$

$$10^{-2} = \frac{1}{10^2} = \tfrac{1}{10} \times \tfrac{1}{10} = \tfrac{1}{100} = .01$$

$$10^{-3} = \frac{1}{10^3} = \tfrac{1}{10} \times \tfrac{1}{10} \times \tfrac{1}{10} = \tfrac{1}{1000} = .001$$

and so on

Multiplying powers of ten. To **multiply powers of ten,** add the exponents.

Example 1: Multiply the following and leave the answers in powers of ten.

(a) 100×10 (d) $10,000 \times .01$

(b) 1000×100 (e) $.0001 \times 1000$

(c) $.01 \times .001$

(a) $$100 \times 10 = 10^2 \times 10^1 = 10^{(2+1)} = 10^3$$

(b) $$1000 \times 100 = 10^3 \times 10^2 = 10^{(3+2)} = 10^5$$

(c) $$.01 \times .001 = 10^{-2} \times 10^{-3} = 10^{[-2+(-3)]} = 10^{-5}$$

(d) $$10,000 \times .01 = 10^4 \times 10^{-2} = 10^{[4+(-2)]} = 10^2$$

(e) $$.0001 \times 1000 = 10^{-4} \times 10^3 = 10^{(-4+3)} = 10^{-1}$$

Dividing powers of ten. To **divide powers of 10,** subtract the exponents; that is, subtract the exponent of the second number (the divisor).

Example 2: Divide the following and leave the answers in powers of ten.

(a) $1000 \div 100$ (d) $.001 \div .01$

(b) $100 \div 10,000$ (e) $10,000 \div .1$

(c) $.1 \div .01$

(a) $$1000 \div 100 = 10^3 \div 10^2 = 10^{(3-2)} = 10^1 \text{ or } 10$$

(b) $\qquad 100 \div 10{,}000 = 10^2 \div 10^4 = 10^{(2-4)} = 10^{-2}$

(c) $\quad 1 \div .01 = 10^{-1} \div 10^{-2} = 10^{[-1-(-2)]} = 10^{(-1+2)} = 10^1$

(d) $\;.001 \div .01 = 10^{-3} \div 10^{-2} = 10^{[-3-(-2)]} = 10^{(-3+2)} = 10^{-1}$

(e) $\quad 10{,}000 \div .1 = 10^4 \div 10^{-1} = 10^{[4-(-1)]} = 10^{(4+1)} = 10^5$

Scientific Notation

Very large or very small numbers are sometimes written in **scientific notation.** A number written in scientific notation is a number between 1 and 10 multiplied by a power of 10.

Example 3: Express the following in scientific notation.

(a) 3,400,000   (c) .0047

(b) .0000008   (d) 27,410

(a) 3,400,000 written in scientific notation is 3.4×10^6. Simply place the decimal point to get a number between 1 and 10 and then count the digits to the right of the decimal to get the power of 10.

$\qquad$ 3.400000.   moved 6 digits to the left

(b) .0000008 written in scientific notation is 8×10^{-7}. Simply place the decimal point to get a number between 1 and 10 and then count the digits from the original decimal point to the new one.

$\qquad$ .0000008.   moved 7 digits to the right

Notice that whole numbers have positive exponents and fractions have negative exponents.

BASIC MATH AND PRE-ALGEBRA

(c) $\qquad$ $.004.7 = 4.7 \times 10^{-3}$

(d) $\qquad$ $2.7430. = 2.743 \times 10^4$

Multiplication in scientific notation. To **multiply** numbers in **scientific notation,** simply multiply the numbers that are between 1 and 10 together to get the first number and add the powers of ten to get the second number.

Example 4: Multiply the following and express the answers in scientific notation.

(a) $(2 \times 10^2)(3 \times 10^4)$ (d) $(5 \times 10^4)(9 \times 10^2)$

(b) $(6 \times 10^5)(5 \times 10^7)$ (e) $(2 \times 10^2)(4 \times 10^4)(5 \times 10^6)$

(c) $(4 \times 10^{-4})(2 \times 10^5)$

(a) $\qquad (2 \times 10^2)(3 \times 10^4) = (2 \times 10^2)(3 \times 10^4)$

$$= 6 \times 10^6$$

(b) $\qquad (6 \times 10^5)(5 \times 10^7) = (6 \times 10^5)(5 \times 10^7)$

$$= 30 \times 10^{12}$$

This answer must be changed to scientific notation (first number from 1 to 9)

$$30 \times 10^{12} = 3.0 \times 10^1 \times 10^{12}$$

$$= 3.0 \times 10^{13}$$

(c) $\qquad (4 \times 10^{-4})(2 \times 10^5) = (4 \times 10^{-4})(2 \times 10^5)$

$$= 8 \times 10^1$$

(d) $(5 \times 10^4)(9 \times 10^2) = (\overbrace{5 \times 10^4})(\overbrace{9 \times 10^2})$

$$= 45 \times 10^6$$

$$= 4.5 \times 10^1 \times 10^6$$

$$= 4.5 \times 10^7$$

(e) $(2 \times 10^2)(4 \times 10^4)(5 \times 10^6) = (2 \times 10^2)(4 \times 10^4)(5 \times 10^6)$

$$= 40 \times 10^{12}$$

$$= 4.0 \times 10^1 \times 10^{12}$$

$$= 4.0 \times 10^{13}$$

Division in scientific notation. To **divide** numbers in **scientific notation,** simply divide the numbers that are between 1 and 10 to get the first number and subtract the powers of ten to get the second number.

Example 5: Divide the following and express the answers in scientific notation.

(a) $(8 \times 10^5) \div (2 \times 10^2)$ (d) $(2 \times 10^4) \div (5 \times 10^2)$

(b) $\dfrac{7 \times 10^9}{4 \times 10^3}$

(e) $(8.4 \times 10^5) \div (2.1 \times 10^{-4})$

(c) $(6 \times 10^7) \div (3 \times 10^9)$

(a) $(8 \times 10^5) \div (2 \times 10^2) = (8 \times 10^5) \div (2 \times 10^2)$

$$= 4 \times 10^3$$

(b) $\dfrac{7 \times 10^9}{4 \times 10^3} = (7 \div 4)(10^9 \div 10^3)$

$$= 1.75 \times 10^6$$

(c) $(6 \times 10^7) \div (3 \times 10^9) = (6 \times 10^7) \div (3 \times 10^9)$

$$= 2 \times 10^{-2}$$

(d) $(2 \times 10^4) \div (5 \times 10^2) = (2 \times 10^4) \div (5 \times 10^2)$

$$= .4 \times 10^2$$

This answer must be changed to scientific notation.

$$.4 \times 10^2 = 4 \times 10^{-1} \times 10^2$$
$$= 4 \times 10^1$$

(e) $(8.4 \times 10^5) \div (2.1 \times 10^{-4}) = (8.4 \times 10^5) \div (2.1 \times 10^{-4})$

$$= 4 \times 10^{5-(-4)}$$
$$= 4 \times 10^9$$

Measurement Systems

English system. The **English system** of measurement is used throughout the United States, although the metric system is being phased in as well. You should be familiar with some basic measurements of the English system. It would be valuable to memorize most of these.

- **Length**

  12 inches (in) = 1 foot (ft)
  3 feet = 1 yard (yd)
  36 inches = 1 yard
  1760 yards = 1 mile (mi)
  5280 feet = 1 mile

- **Area**

  144 square inches (sq in) = 1 square foot (sq ft)
  9 square feet = 1 square yard (sq yd)

- **Weight**

  16 ounces (oz) = 1 pound (lb)
  2000 pounds = 1 ton (T)

- **Capacity**

  2 cups = 1 pint (pt)
  2 pints = 1 quart (qt)
  4 quarts = 1 gallon (gal)
  4 pecks = 1 bushel

■ **Time**

365 days = 1 year
52 weeks = 1 year
10 years = 1 decade
100 years = 1 century

Metric system. The **metric system** of measurement is based on powers of ten. To understand the metric system, it is important to know the meaning of the prefixes to each base unit.

■ **Prefixes**

kilo = thousand
hecto = hundred
deka = ten

deci = tenth
centi = hundredth
milli = thousandth

Now, applying these to length, volume, and mass should be easier.

■ **Length—meter**

1 kilometer (km) = 1000 meters (m)
1 hectometer (hm) = 100 meters
1 dekameter (dam) = 10 meters

10 decimeters (dm) = 1 meter
100 centimeters (cm) = 1 meter
1000 millimeters (mm) = 1 meter

■ **Volume—liter**

1000 milliliters (ml, or mL) = 1 liter (l, or L)
1000 liters = 1 kiloliter (kl, or kL)

- **Mass—gram**

  1000 milligrams (mg) = 1 gram (g)
  1000 grams = 1 kilogram (kg)
  1000 kilograms = 1 metric ton (t)

- **Some approximations**

  a meter is a little more than a yard
  a kilometer is about .6 mile
  a kilogram is about 2.2 pounds
  a liter is slightly more than a quart

Converting Units of Measure

Example 1: If 36 inches equal 1 yard, then 3 yards equal how many inches?

Intuitively, $3 \times 36 = 108$ in

By proportion, using yards over inches,

$$\frac{3}{x} = \frac{1}{36}$$

Remember to place the same units across from each other—inches across from inches, etc. Then solve.

$$\frac{3}{x} = \frac{1}{36}$$

Cross multiply.

$$108 = x$$

$$x = 108 \text{ in}$$

Example 2: If 2.2 pounds equal 1 kilogram, then 10 pounds equal approximately how many kilograms?

Intuitively, $10 \div 2.2 = 4.5 \text{ kg}$

By proportion, using kilograms over pounds,

$$\frac{x}{10} = \frac{1}{2.2}$$

Cross multiply.

$$2.2x = 10$$

Divide and cancel.

$$\frac{2.2x}{2.2} = \frac{10}{2.2}$$

$$x = 4.5 \text{ kg}$$

Example 3: Change 3 decades into weeks.

Since 1 decade equals 10 years, and 1 year equals 52 weeks, then 3 decades equal 30 years.

30 years × 52 weeks = 1560 weeks in 30 years, or 3 decades

Notice that this was converted step by step. It could have been done in one step.

$$3 \times 10 \times 52 = 1560 \text{ weeks}$$

Example 4: If 1760 yards equal 1 mile, how many yards are in 5 miles?

$$1760 \times 5 = 8800 \text{ yards in 5 miles}$$

Example 5: If 1 kilometer equals 1000 meters, and 1 dekameter equals 10 meters, how many dekameters are in 3 kilometers?

$$\frac{1}{10} = \frac{x}{3000}$$

$$10x = 3000$$

$$\frac{\cancel{10}x}{\cancel{10}} = \frac{3000}{10}$$

$$x = 300 \text{ dam}$$

Precision

The word *precision* refers to the degree of exactness of a measurement, that is, how *fine* the measurement is. The smaller the unit of measure used, the more **precise** is the measurement. Keep in mind that the precision of a measurement has nothing to do with the size of the numbers, only with the unit used.

Example 6: Which of the following measurements is more precise?

(a) 5.4 mm   or   3.22 mm   (d) 3.67 m   or   5.1 m

(b) 1 m   or   1 km   (e) 5.69 cm   or   9.99 cm

(c) 3 ft   or   11 in

(a)   Since the smallest unit in 5.4 mm is .4, and the smallest unit in 3.22 mm is .02, 3.22 mm is more precise.

(b)   Since meter (m) is a smaller unit than kilometer (km), 1 m is more precise

(c)   Since inch (in) is a smaller unit that foot (ft), 11 inches is more precise

(d) Since 3.67 goes to the hundredths, and 5.1 goes to the tenths, 3.67 is more precise.

(e) Since each one goes to the hundredths place, they have the same precision.

When adding or subtracting two measures, you *cannot* be more precise than the least precise unit being used. In other words, the unit being used in the answer should be the same as the less precise of the units used in the two measurements.

Example 7: Perform the indicated operation and give the answer in the less precise unit.

(a) $5.44 \text{ km} + 2.1 \text{ km}$ (b) $32.77 \text{ g} - 12 \text{ g}$

First round off the more precise number and then calculate

(a)
$$
\begin{array}{r}
5.44 \text{ km} \approx 5.4 \text{ km} \\
+\ 2.1 \ \ \text{km} = 2.1 \text{ km} \\
\hline
7.5 \text{ km}
\end{array}
$$

(b)
$$
\begin{array}{r}
32.77 \text{ g} \approx 33 \text{ g} \\
-\ 12 \ \ \ \ \text{g} = 12 \text{ g} \\
\hline
21 \text{ g}
\end{array}
$$

Significant Digits

When a digit tells how many units of measure are involved, it is a **significant digit.** To find the number of units of measure, simply divide the actual measurement by the unit of measure.

Example 8: Find the number of units used and the significant digits for each of the following.

(a) 23.7 m (unit of measure, 0.1 m)

(b) .05 cm (unit of measure, 0.01 cm)

(c) 520 g (unit of measure, 1 g)

(d) 760 km (unit of measure, 10 km)

(e) 2100 m (unit of measure, 10 m)

(f) 2100 m (unit of measure, 100 m)

For each of these, find the number of units of measure by dividing the actual measurement by the unit of measure. For example, for (a),

$$0.1\overline{)23.7} = 237$$

	Measure-ment	Unit of Measure	Number of Units Used	Significant Digits
(a)	23.7 m	0.1 m	237 units of 0.1 m	237
(b)	.05 cm	0.01 cm	5 units of .01 cm	5
(c)	520 g	1 g	520 units of 1 g	520
(d)	760 km	10 km	76 units of 10 km	76
(e)	2100 m	10 m	210 units of 10 m	210
(f)	2100 m	100 m	21 units of 100 m	21

The following rules can be used as general guidelines for determining significant digits.

1. Digits that are not zero are always significant. In 35.7, all three digits are significant.

2. Zeros at the end (or right) of the decimal are always significant. In .20, both digits are significant.

3. In a decimal, zeros in front (or to the left) of a significant digit are never significant. In .007, only 7 is significant.

4. All zeros that are between significant digits are significant. In 500.5, all digits are significant.

5. Depending on the unit of mesure, zeros that are at the end (or right) of a whole number may or may not be significant. For example, 500 m measured to the 1 m gives three significant digits (5, 0, and 0), while 500 m measured to the 10 m gives two significant digits (5 and 0).

When computing using significant digits, you should always round the answer to the smallest number of significant digits that is in any of the numbers being used.

Example 9: Find the area of the rectangle with length 3.1 m and width 2.2 m.

$$
\begin{array}{ll}
3.1 & \text{(2 significant digits)} \\
\times\ 2.2 & \text{(2 significant digits)} \\
\hline
6.82 & \text{(3 significant digits must be changed to 2 significant digits)}
\end{array}
$$

So
$$6.82 \approx 6.8 \text{ sq m}$$

Example 10: Find the length of a rectangle with area 19 sq cm and width 4 cm.

Because 19 has two significant digits and 4 has one significant digit, the answer must be rounded to one significant digit.

$$4\overline{)19.00} = 4.75$$

So $$4.75 \approx 5 \text{ cm}$$

Calculating Measurements of Basic Figures

Some basic figures, such as squares, rectangles, parallelograms, trapezoids, triangles, and circles, have measurements that are not difficult to calculate if the necessary information is given and the proper formula is used. You should first be familiar with each of the following formulas.

Perimeter of some polygons—squares, rectangles, parallelograms, trapezoids, and triangles. Perimeter (*P*) means the total distance all the way around the outside of the **polygon** (a many-sided plane closed figure). The perimeter of that polygon can be determined by adding up the lengths of all the sides. The total distance around will be the sum of all sides of the polygon. No special formulas are really necessary, although two are commonly seen.

- **perimeter (*P*) of a square** = $4s$ (s = length of side)

- **perimeter (*P*) of a parallelogram** (rectangle and rhombus) = $2l + 2w$ or $2(l + w)$ (l = length, w = width)

Area of some polygons—squares, rectangles, parallelograms, trapezoids, and triangles. Area (A) means the amount of space inside the polygon. The formulas for each area are as follows.

■ **triangle (Figure 3):** $A = \frac{1}{2}bh$ (b = base, h = height)

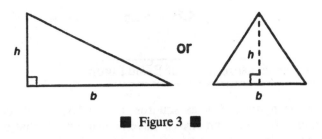

■ Figure 3 ■

A triangle is a three-sided polygon. In a triangle, the base is the side the triangle is resting on, and the height is the distance from the base to the opposite point, or **vertex.**

Example 11: What is the area of the triangle shown in Figure 4?

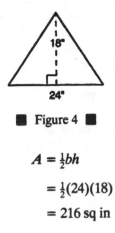

■ Figure 4 ■

$$A = \frac{1}{2}bh$$

$$= \frac{1}{2}(24)(18)$$

$$= 216 \text{ sq in}$$

- **square or rectangle** (Figure 5): $A = lw$

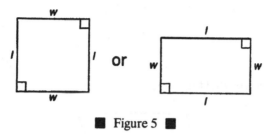

■ Figure 5 ■

A square is a four-sided polygon with all sides equal and all right angles (90 degrees). A rectangle is a four-sided polygon with opposites sides equal and all right angles. In a square or rectangle, the bottom, or resting side, is the base, and either adjacent side is the height.

Example 12: What is the area of these polygons?

(a) the square shown in Figure 6(a)

(b) the rectangle shown in Figure 6(b)

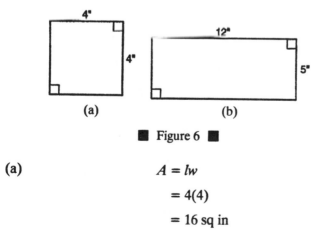

■ Figure 6 ■

(a)
$$A = lw$$
$$= 4(4)$$
$$= 16 \text{ sq in}$$

(b)
$$A = lw$$
$$= 12(5)$$
$$= 60 \text{ sq in}$$

- **parallelogram** (Figure 7): $A = bh$

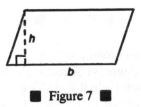

■ Figure 7 ■

A parallelogram is a four-sided polygon with opposite sides equal. In a parallelogram, the resting side is usually considered the base, and a perpendicular line going from the base to the other side is the height.

Example 13: What is the area of the parallelogram shown in Figure 8?

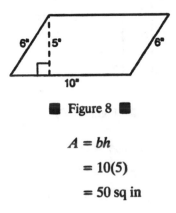

■ Figure 8 ■

$$A = bh$$
$$= 10(5)$$
$$= 50 \text{ sq in}$$

- **trapezoid** (Figure 9): $A = \frac{1}{2}(b_1 + b_2)h$

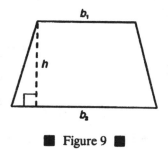

■ Figure 9 ■

A trapezoid is a four-sided polygon with only two sides parallel. In a trapezoid, the parallel sides are the bases, and the distance between the two bases is the height.

Example 14: What is the area of the trapezoid shown in Figure 10?

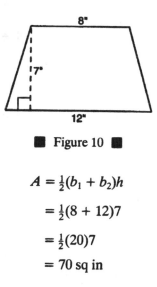

■ Figure 10 ■

$$A = \frac{1}{2}(b_1 + b_2)h$$
$$= \frac{1}{2}(8 + 12)7$$
$$= \frac{1}{2}(20)7$$
$$= 70 \text{ sq in}$$

Example 15: What is the perimeter (P) and the area (A) of the polygons shown in Figure 11, (a) through (f), in which all measures shown are in inches?

(a) $P = 15 + 13 + 14$

    $= 42$ in

$A = \frac{1}{2}bh$

    $= \frac{1}{2}(14)(12)$

    $= 84$ sq in

(b) $P = 6 + 8 + 10$

    $= 24$ in

$A = \frac{1}{2}bh$

    $= \frac{1}{2}(8)(6)$

    $= 24$ sq in

(c) $P = 10 + 10 + 2 + 2$

    $= 24$ in

$A = bh$

    $= 10(2)$

    $= 20$ sq in

(d) $P = 10 + 10 + 5 + 5$

    $= 30$ in

$A = bh$

    $= 10(4)$

    $= 40$ sq in

(e) $P = 5 + 5 + 5 + 5$

    $= 20$ in

$A = bh$

    $= 4(5)$

    $= 20$ sq in

(f) $P = 17 + 7 + 10 + 28$

    $= 62$ in

$A = \frac{1}{2}(b_1 + b_2)h$

    $= \frac{1}{2}(7 + 28)(8)$

    $= 4(35)$

    $= 140$ sq in

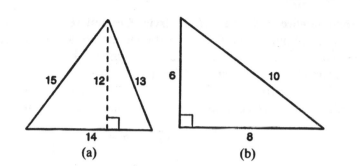

(a) (b)

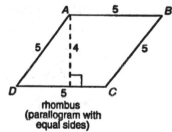

rectangle *ABCD*
(c)

parallelogram *ABCD*
(d)

rhombus
(parallelogram with
equal sides)
(e)

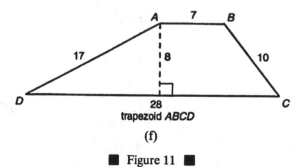

trapezoid *ABCD*
(f)

■ Figure 11 ■

Circumference and area of a circle. Circumference (*C*) is the distance around the circle. Since the circumference of any circle divided by its **diameter** (*d*) (the line segment that contains the center and has its end points on the circle) yields the same value, the Greek letter π (pi) is used to represent that value. In decimal or fractional form, the commonly used approximations of π are

$$\pi \approx 3.14 \quad \text{or} \quad \pi \approx \tfrac{22}{7}$$

Use either value in your calculations. The formula for circumference is

$$C = \pi d \quad \text{or} \quad C = 2\pi r$$

(*r* = **radius,** a line segment whose endpoints lie one at the center of the circle and one on the circle, half the length of the diameter)

Example 16: What is the circumference of circle *M* shown in Figure 12?

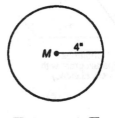

■ Figure 12 ■

In circle *M*, *d* = 8, since *r* = 4.

$$C = \pi d$$
$$= \pi(8)$$
$$\approx 3.14(8)$$
$$\approx 25.12 \text{ in}$$

The **area** (*A*) of a circle can be determined by

$$A = \pi r^2$$

Example 17: What is the area of circle *M* shown in Figure 13?

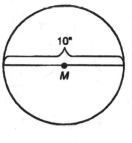

■ Figure 13 ■

In circle *M*, *r* = 5, since *d* = 10.

$$
\begin{aligned}
A &= \pi r^2 \\
&= \pi(5^2) \\
&\approx 3.14(25) \\
&\approx 78.5 \text{ sq in}
\end{aligned}
$$

Example 18: From the given radius or diameter, find the area and circumference (leave in terms of π) of the circles shown in Figure 14.

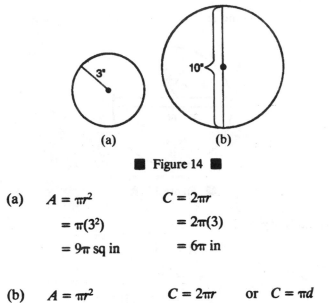

(a)   (b)

■ Figure 14 ■

(a) $A = \pi r^2$ $C = 2\pi r$

   $\quad = \pi(3^2)$ $\quad = 2\pi(3)$

   $\quad = 9\pi$ sq in $\quad = 6\pi$ in

(b) $A = \pi r^2$ $C = 2\pi r$ or $C = \pi d$

   $\quad = \pi(5^2)$ $\quad = 2\pi(5)$ $\quad = 10\pi$ in

   $\quad = 25\pi$ sq in $\quad = 10\pi$ in

Information may be displayed in many ways. The three basic types of graphs you should know are **bar graphs, line graphs,** and **pie graphs** (or **pie charts**).

When answering questions related to a graph, you should

1. Examine the entire graph—notice labels and headings.
2. Focus on the information given.
3. Look for major changes—high points, low points, trends.
4. Don't memorize the graph, refer to it.
5. Pay special attention to which part of the graph the question is referring to.
6. If you don't understand the graph, reread the headings and labels.

Bar Graphs

Bar graphs convert the information in a chart into separate bars or columns. Some graphs list numbers along one edge and places, dates, people, or things (individual categories) along another edge. Always try to determine the *relationship* between the columns in a graph or chart.

Example 1: The bar graph shown in Figure 15 indicates that City W has approximately how many more billboards than does City Y?

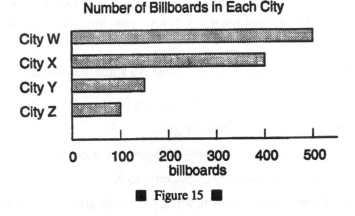

Number of Billboards in Each City

■ Figure 15 ■

Notice that the graph shows the "Number of Billboards in Each City," with the numbers given along the bottom of the graph in increases of 100. The names are listed along the left side. City W has approximately 500 billboards. The bar graph for City Y stops about halfway between 100 and 200. Now, consider that halfway between 100 and 200 would be 150. So City W (500) has approximately 350 more billboards than does City Y (150).

$$500 - 150 = 350$$

Example 2: Based on the bar graph shown in Figure 16.

(a) The number of books sold by Mystery Mystery 1990–92 exceeded the number of those sold by All Sports by approximately how many?

(b) From 1991 to 1992, the percent increase in number of books sold for All Sports exceeded the percent increase of Mystery Mystery by approximately how much?

CLIFFS QUICK REVIEW

(c) What caused the 1992 decline in Reference Unlimited's number of books sold?

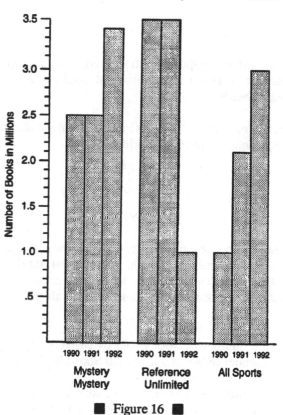

Number of Books Sold by Three Publishers

■ Figure 16 ■

This graph contains multiple bars representing each publisher. Each single bar stands for the number of books sold in a single year. You may be tempted to write out the numbers as you do your arithmetic (3.5 million = 3,500,000). This is unnecessary, as it often is on graphs which use large numbers. Since *all* measurements are in millions, adding zeros does not add precision to the numbers.

(a) Referring to the Mystery Mystery bars, you can see that number of books sold per year is as follows.

$$1990 = 2.5$$
$$1991 = 2.5$$
$$1992 = 3.4$$

Use a piece of paper as a straightedge to determine this last number. Totaling the number of books sold for all three years gives 8.4.

Referring to the All Sports bars, you can see that number of books sold per year is as follows.

$$1990 = 1$$
$$1991 = 2.1$$
$$1992 = 3$$

Once again, use a piece of paper as a straightedge, but don't designate numbers beyond the nearest tenth, since the graph numbers prescribe no greater accuracy than this. Totaling the number of books sold for all three years gives 6.1.

So the number of books sold by Mystery Mystery exceeds the number of books sold by All Sports by 2.3 million.

(b) Graph and chart questions may ask you to calculate percent increase or percent decrease. As you learned in the section dealing with percents, the formula for figuring either of these is the same.

$$\frac{\text{change}}{\text{starting point}} = \text{percent change}$$

In this case, the percent increase in number of books sold by Mystery Mystery can be calculated first.

number of books sold in 1991 = 2.5

number of books sold in 1992 = 3.4

change = .9

The 1991 amount is the "starting point," so

$$\frac{\text{change}}{\text{starting point}} = \frac{.9}{2.5} = .36 = 36\%$$

The percent increase in number of books sold by All Sports can be calculated as follows.

number of books sold in 1991 = 2.1

number of books sold in 1992 = 3

change = .9

$$\frac{\text{change}}{\text{starting point}} = \frac{.9}{2.1} = .428 \approx 43\%$$

So the percent increase of All Sports exceeds that of Mystery Mystery by 7%.

$$43\% - 36\% = 7\%$$

(c) This question cannot be answered based on the information in the graph. Never assume information not given. In this case, the multiple factors which could cause a decline in number of books sold are not represented by the graph.

Line Graphs

Line graphs convert data into points on a grid. These points are then connected to show a relationship among the items, dates, times, etc. Notice the slopes of the lines connecting the points. These lines will show increases and decreases. The sharper the slope *upward*, the greater the *increase*. The sharper the slope *downward*, the greater the *decrease*. Line graphs can show trends, or changes, in data over a period of time.

Example 3: Based on the line graph shown in Figure 17,

(a) In what year was the property value of Moose Lake Resort about $500,000?

(b) In which ten-year period was there the greatest decrease in the property value of Moose Lake Resort?

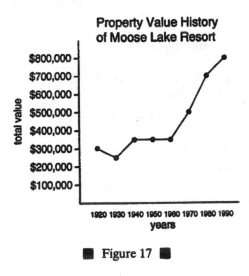

■ Figure 17 ■

(a) The information along the left side of the graph shows the property value of Moose Lake Resort in increments of $100,000. The bottom of the graph shows the years from 1920 to 1990. You will notice that in 1970 the property value was about $500,000. Using the edge of a sheet of paper as a ruler will help you see that the dot in the 1970 column lines up with $500,000 on the left.

(b) Since the slope of the line goes *down* from 1920 to 1930, there must have been a decrease in property value. If you read the actual numbers, you will notice a decrease from $300,000 to about $250,000.

Example 4: According to the line graph shown in Figure 18, the tomato plant grew the most between which two weeks?

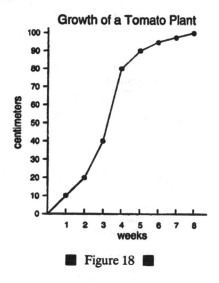

Growth of a Tomato Plant

■ Figure 18 ■

The numbers at the bottom of the graph give the weeks of growth of the plant. The numbers on the left give the height of the plant in centimeters. The sharpest upward slope occurs between week 3 and week 4, when the plant grew from 40 centimeters to 80 centimeters, a total of 40 centimeters growth.

Circle Graphs, or Pie Charts

A **circle graph,** or **pie chart,** shows the relationship between the whole circle (100%) and the various slices that represent portions of that 100%. The larger the slice, the higher the percentage.

Example 5: Based on the circle graph shown in Figure 19,

   (a) If Smithville Community Theater has $1000 to spend this month, how much will be spent on set construction?

   (b) What is the ratio of the amount of money spent on advertising to the amount of money spent on set construction?

Figure 19

   (a) The theater spends 20% of its money on set construction. 20% of $1000 is $200, so $200 will be spent on set construction.

(b) To answer this question, you must use the information in the graph to make a ratio.

$$\frac{\text{advertising}}{\text{set construction}} = \frac{15\% \text{ of } 1000}{20\% \text{ of } 1000} = \frac{150}{200} = \frac{3}{4}$$

Notice that $\frac{15\%}{20\%}$ reduces to $\frac{3}{4}$.

Example 6: Based on the circle graph shown in Figure 20,

(a) If the Bell Canyon PTA spends the same percentage on dances every year, how much will they spend on dances in a year in which their total amount spent is $15,000?

(b) The amount of money spent on field trips in 1995 was approximately what percent of the total amount spent?

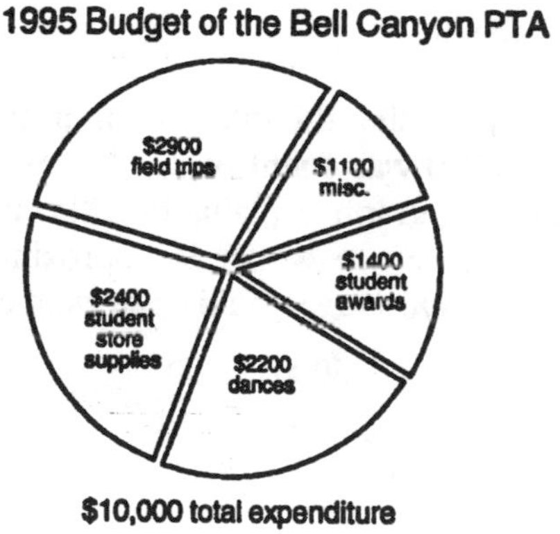

1995 Budget of the Bell Canyon PTA

$10,000 total expenditure

■ Figure 20 ■

(a) To answer this question, you must find a percent and then apply this percent to a new total. In 1995, the PTA spent $2200 on dances. This can be calculated to be 22% of the total spent in 1995 by the following method.

$$\frac{2200}{10,000} = \frac{22}{100} = 22\%$$

Now, multiplying 22% times the *new* total amount spent of $15,000 will give the right answer.

$$22\% = .22$$

$$.22 \times 15,000 = 3300 \quad \text{or} \quad \$3300$$

You could use another common-sense method. If $2200 out of $10,000 is spent for dances, $1100 out of every $5000 is spent for dances. Since $15,000 is 3 × $5000, 3 × $1100 would be $3300.

(b) By carefully reading the information in the graph, you will find that $2900 was spent on field trips. The information describing the graph explains that the total expenditures were $10,000. Since $2900 is approximately $3000, the approximate *percentage* would be worked out as follows.

$$\frac{3000}{10,000} = \frac{30}{100} = 30\%$$

Coordinate Graphs

Each point on a number line is assigned a number. In the same way, each point in a plane is assigned a pair of numbers. These numbers represent the placement of the point relative to two intersecting lines. In **coordinate graphs** (Figure 21), two perpendicular number lines are used and are called **coordinate axes**. One axis is horizontal and is called the **x-axis**. The other is vertical and is called the **y-axis**. The point of intersection of the two number lines is called the **origin** and is represented by the coordinates (0, 0).

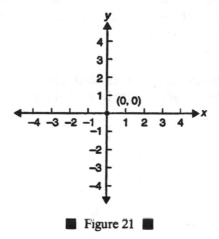

■ Figure 21 ■

Each point on a plane is located by a unique ordered pair of numbers called the **coordinates**. Some coordinates are noted in Figure 22.

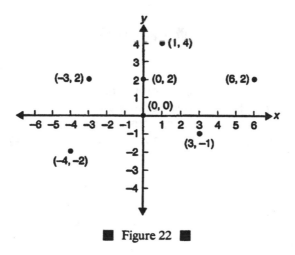

■ Figure 22 ■

Notice that on the *x*-axis numbers to the right of 0 are positive and to the left of 0 are negative. On the *y*-axis, numbers above 0 are positive and below 0 are negative. Also note that the first number in the ordered pair is called the **x-coordinate,** or **abscissa,** while the second number is the **y-coordinate,** or **ordinate.** The *x*-coordinate shows the right or left direction, and the *y*-coordinate shows the up or down direction.

The coordinate graph is divided into four quarters called **quadrants.** These quadrants are labeled in Figure 23.

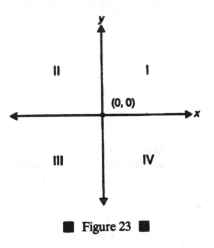

■ Figure 23 ■

Notice that

In quadrant I, *x* is always positive and *y* is always positive.
In quadrant II, *x* is always negative and *y* is always positive.
In quadrant III, *x* and *y* are both always negative.
In quadrant IV, *x* is always positive and *y* is always negative.

Example 7: Identify the points (*A, B, C, D, E,* and *F*) on the coordinate graph shown in Figure 24.

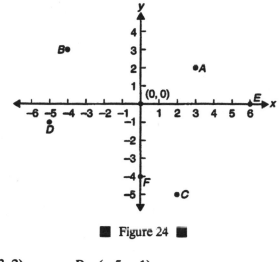

■ Figure 24 ■

A	(3, 2)	*D*	(−5, −1)
B	(−4, 3)	*E*	(6, 0)
C	(2, −5)	*F*	(0, −4)

Probability

Probability is the numerical measure of the chance of an outcome or event occurring. **When all outcomes are equally likely to occur,** the probability of the occurrence of a given outcome can be found by using the following formula.

$$\text{probability} = \frac{\text{number of favorable outcomes}}{\text{number of possible outcomes}}$$

Example 1: Using the spinner shown in Figure 25, what is the probability of spinning a 6 in one spin?

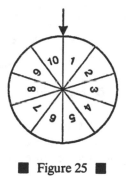

■ Figure 25 ■

Since there is only *one* 6 on the spinner out of *ten* numbers and all the numbers are equally spaced, the probability is $\frac{1}{10}$.

Example 2: Once again using the spinner shown in Figure 25, what is the probability of spinning either a 3 or a 5 in one spin?

Since there are *two favorable outcomes* out of *ten possible outcomes*, the probability is $\frac{2}{10}$, or $\frac{1}{5}$.

When two events are independent of each other, you need to multiply to find the favorable and/or possible outcomes.

Example 3: What is the probability that both of the spinners shown in Figure 26 will stop on a 3 on the first spin?

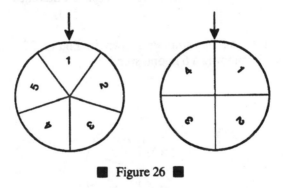

■ Figure 26 ■

Since the probability that the first spinner will stop on the number 3 is $\frac{1}{5}$ and the probability that the second spinner will stop on the number 3 is $\frac{1}{4}$, and since each event is independent of the other, simply multiply.

$$\tfrac{1}{5} \times \tfrac{1}{4} = \tfrac{1}{20}$$

Example 4: What is the probability that on two consecutive rolls of a die the numbers will be 2 and then 3?

Since the probability of getting a 2 on the first roll is $\frac{1}{6}$ and the probability of getting a 3 on the second roll is $\frac{1}{6}$, and since the rolls are independent of each other, simply multiply.

$$\frac{1}{6} \times \frac{1}{6} = \frac{1}{36}$$

Example 5: What is the probability of tossing heads three consecutive times with a two-sided fair coin?

Since each toss is independent and the odds are $\frac{1}{2}$ for each toss, the probability would be

$$\frac{1}{2} \times \frac{1}{2} \times \frac{1}{2} = \frac{1}{8}$$

Example 6: What is the probability of rolling two dice in one toss so that they total 5?

Since there are six possible outcomes on each die, the total possible outcomes for two dice is

$$6 \times 6 = 36$$

The favorable outcomes are $(1 + 4)$, $(4 + 1)$, $(2 + 3)$, and $(3 + 2)$. These are all the ways of tossing a total of 5 on two dice. Thus, there are four favorable outcomes, which gives the probability of throwing a five as

$$\frac{4}{36} = \frac{1}{9}$$

Example 7: Three green marbles, two blue marbles, and five yellow marbles are placed in a jar. What is the probability of selecting at random a green marble on the first draw?

Since there are ten marbles (total possible outcomes) and three green marbles (favorable outcomes), the probability is $\frac{3}{10}$.

Example 8: In a regular deck of fifty-two cards, what is the probability of drawing a heart on the first draw? (There are thirteen hearts in a deck.)

Since there are thirteen favorable outcomes out of fifty-two possible outcomes, the probability is $\frac{13}{52}$, or $\frac{1}{4}$.

Combinations. If there are a **number of successive choices** to make and the choices are **independent of each other** (order makes no difference), the total number of possible choices **(combinations)** is the product of each of the choices at each stage.

Example 9: How many possible combinations of shirts and ties are there if there are five different color shirts and three different color ties?

To find the total number of possible combinations, simply multiply the number of shirts times the number of ties.

$$5 \times 3 = 15$$

Example 10: A combination lock has three settings, each of which contains numbers from 0 to 9. How many different possible combinations exist on the lock?

Note that each setting is independent of the others; thus, since each has ten possible settings,

$$10 \times 10 \times 10 = 1000$$

There are 1000 possible combinations.

Permutations. If there are a **number of successive choices** to make and the choices are **affected by the previous choice or choices** (dependent upon order), then **permutations** are involved.

Example 11: How many ways can you arrange the letters S, T, O, P in a row?

number of choices for first letter		*number of choices for second letter*		*number of choices for third letter*		*number of choices for fourth letter*
4	$\times$	3	$\times$	2	$\times$	1

$$4! = 4 \times 3 \times 2 \times 1 = 24$$

The product $4 \times 3 \times 2 \times 1$ can be written 4! (read *4 factorial* or *factorial 4*). Thus, there are twenty-four different ways to arrange four different letters.

Example 12: How many different ways are there to arrange three jars in a row on a shelf?

Since the order of the items is affected by the previous choice(s), the number of different ways equals 3!, or

$$3 \times 2 \times 1 = 6$$

There are six different ways to arrange the three jars.

Example 13: If, from among five people, three executives are to be selected, how many possible combinations of executives are there?

This is a more difficult type of combination involving permutations. Notice here that the order of selection makes no difference. The symbol used to denote this situation is $C(n, r)$, which is read *the number of combinations of n things taken r at a time*. The formula used is

$$C(n, r) = \frac{n!}{r!(n - r)!}$$

Since $n = 5$ and $r = 3$ (five people taken three at a time), then the solution is as follows:

$$\frac{5!}{3!(5 - 3)!}$$

Now, solve.

$$\frac{5 \cdot 4 \cdot 3 \cdot 2 \cdot 1}{3 \cdot 2 \cdot 1 \, (2)!} = \frac{5 \cdot \overset{2}{\cancel{4}} \cdot \cancel{3} \cdot \cancel{2} \cdot 1}{\cancel{3} \cdot \cancel{2} \cdot 1 \, (\cancel{2} \cdot 1)} = 10$$

If the problem involves very few possibilities, you may wish to actually list the possible combinations.

Example 14: A coach is selecting a starting lineup for her basketball team. She must select from among nine players to get her starting lineup of five. How many possible starting lineups could she have?

Since $n = 9$ and $r = 5$ (nine players taken five at a time), the solution is as follows.

$$\frac{9!}{5!(9 - 5)!}$$

$$\frac{9 \cdot 8 \cdot 7 \cdot 6 \cdot 5 \cdot 4 \cdot 3 \cdot 2 \cdot 1}{5 \cdot 4 \cdot 3 \cdot 2 \cdot 1 \, (4)!} = \frac{9 \cdot \overset{2}{\cancel{8}} \cdot 7 \cdot \overset{1}{\cancel{6}} \cdot \cancel{5} \cdot \cancel{4} \cdot \cancel{3} \cdot \cancel{2} \cdot 1}{\cancel{5} \cdot \cancel{4} \cdot \cancel{3} \cdot \cancel{2} \cdot 1 \, (\cancel{4} \cdot \cancel{3} \cdot \cancel{2} \cdot 1)} = 126$$

Example 15: How many possible combinations of *a, b, c,* and *d* taken two at a time are there?

Since $n = 4$ and $r = 2$ (four letters taken two at a time), the solution is as follows.

$$\frac{4!}{2!(4-2)!} = \frac{4!}{2!(2)!} = \frac{\overset{2}{\cancel{4}} \cdot 3 \cdot \cancel{2} \cdot 1}{\cancel{2} \cdot 1 \, (\cancel{2} \cdot 1)} = 6$$

You might simply have listed the possible combinations as *ab, ac, ad, bc, bd,* and *cd.*

Statistics

The study of numerical data and their distribution is called **statistics.**

Measures of central tendencies. Any measure indicating a center of a distribution is called a **measure of central tendency.** The three basic measures of central tendency are

1. **mean** (or **arithmetic mean**)
2. **median**
3. **mode**

The **mean (arithmetic mean)** is what is usually called the average. The arithmetic mean is the most frequently used measure of central tendency. It is generally reliable, easy to use, and more stable than the median. To determine the arithmetic mean, simply total the items and then divide by the number of items.

Example 16: What is the arithmetic mean of 0, 12, 18, 20, 31, and 45?

Total the items.

$$0 + 12 + 18 + 20 + 31 + 45 = 126$$

Divide by the number of items

$$126 \div 6 = 21$$

The arithmetic mean is 21.

Example 17: What is the arithmetic mean of 25, 27, 27, and 27?

$$25 + 27 + 27 + 27 = 106$$

$$106 \div 4 = 26\tfrac{1}{2}$$

The arithmetic mean is $26\tfrac{1}{2}$.

Example 18: What is the arithmetic mean of 20 and -10?

$$20 + (-10) = +10$$

$$10 \div 2 = 5$$

The arithmetic mean is 5.

When one or a number of items is used several times, those items have more "weight." This establishment of relative importance, or weighting, is used to compute the **weighted mean.**

Example 19: What is the mean of three tests averaging 70% plus seven tests averaging 85%?

In effect, you have here ten exams, three of which score 70% and seven of which score 85%. Rather than adding all ten scores, to determine the above "weighted mean," simply multiply 3 times 70% to find the total of those items (210). Then multiply 7 times 85% to find their total (595). Now add the two totals (805) and divide by the number of items overall (10). The weighted mean is thus 80.5%.

Example 20: For the first nine months of the year, the average monthly rainfall was two inches. For the last three months of that year, rainfall averaged four inches per month. What was the mean monthly rainfall for the entire year?

$9 \times 2'' = 18''$

$\underline{3 \times 4'' = 12''}$

total $= 30''$ divided by 12 months in all $= 2.5''$ monthly mean

Example 21: Six students averaged 90% on a class test. Four other students averaged 70% on the test. What was the mean score of all ten students?

$6 \times 90 = 540$

$\underline{4 \times 70 = 280}$

total $= 820$ divided by 10 students $= 82\%$

The **median** of a set of numbers arranged in ascending or descending order is the middle number (if there is an odd number of items in the set). If there is an even number of items in the set, their median is the arithmetic mean of the middle two numbers. The median is easy to calculate and is not influenced by extreme measurements.

Example 22: Find the median of 3, 4, 6, 9, 21, 24, 56.

$$3, 4, 6, \underline{9}, 21, 24, 56$$

The median is 9.

Example 23: Find the median of 4, 5, 6, 10.

$$5\tfrac{1}{2}$$
$$4, 5, \quad 6, 10$$

The median is $5\tfrac{1}{2}$.

The set, class, or classes that appear most, or whose frequency is the greatest, is the **mode** or **modal class**. (Mode is not greatly influenced by extreme cases but is probably the least important or least used of the three types.)

Example 24: Find the mode of 3, 4, 8, 9, 9, 2, 6, 11.

The mode is 9 because it appears more often than any other number.

A **number series** is a mathematical progression of numbers or a sequence of numbers with some pattern. And although some series are simply patterns (2, 2, 4, 5, 2, 2, 6, 7, 2, 2, 8, 9, ...), the two common types of mathematical series are **arithmetic** and **geometric progressions**.

Arithmetic Progressions

An **arithmetic progression** is a progression in which there is a common difference between terms. Subtracting any term from the next results in the same value.

Example 1: What is the next number in the progression 4, 7, 10, 13, ... ?

Since the common difference is 3, the next number would be 16.

Example 2: What is the fortieth term in the progression 5, 10, 15, 20, ... ?

The common difference is $10 - 5 = 5$. The fortieth term is 36 terms beyond the fourth term, 20.

$$5 \times 36 = 180$$

$$+ 180$$
$$5, 10, 15, 20, \ldots, 200$$

So the fortieth term is 200.

Geometric Progressions

A **geometric progression** is a progression in which there is a constant ratio between terms. The value found by dividing or multiplying any term by the preceding term will give you the constant ratio.

Example 3: What is the next number in the progression 1, 2, 4, 8, 16, . . .?

The ratio is $\frac{2}{1}$, since you're simply doubling each number to get the next number. So doubling the last number, 16, gives 32.

Example 4: What is the next number in the progression 3, 9, 27, 81, . . .?

The ratio is $\frac{3}{1}$ because you are multiplying each number by 3 to get the next number. So multiply the last number, 81, by 3, which gives 243.

Variables and Algebraic Expressions

Variables. A **variable** is a letter used to stand for a number. The letters x, y, z, a, b, c, m, and n are probably the most commonly used variables. The letters e and i have special values in algebra and are usually not used as variables. The letter o is usually not used because it can be mistaken for 0 (zero).

Algebraic expressions. Variables are used to change verbal expressions into **algebraic expressions,** that is, expressions that are composed of letters that stand for numbers. Key words that will help you translate words into letters and numbers include: for addition, *some, more than, greater than,* and *increase;* for subtraction, *minus, less than, smaller than,* and *decrease;* for multiplication, *times, product,* and *multiplied by;* for division, *halve, divided by,* and *ratio.* Also see "Key Words" in the section "Word Problems."

Example 1: Give the algebraic expression for each of the following.

   (a)   the sum of a number and 5

   (b)   the number minus 4

   (c)   six times a number

   (d) x divided by 7

   (e)   three more than the product of 2 and x

   (a)   The sum of a number and 5: $x + 5$ or $5 + x$.

   (b)   The number minus 4: $x - 4$.

   (c)   Six times a number: $6x$

(d) $\qquad$ x divided by 7: $x/7$ or $\dfrac{x}{7}$.

(e) Three more than the product of 2 and x: $2x + 3$.

Evaluating expressions. To **evaluate an expression,** just replace the variables with grouping symbols, insert the values given for the variables, and do the arithmetic. Remember to follow the order of operations: parentheses, exponents, multiplication/division, addition/subtraction.

Example 2: Evaluate each of the following.

(a) $x + 2y$ if $x = 2$ and $y = 5$

(b) $a + bc - 3$ if $a = 4$, $b = 5$, and $c = 6$

(c) $m^2 + 4n + 1$ if $m = 3$ and $n = 2$

(d) $\dfrac{b + c}{7} + \dfrac{c}{a}$ if $a = 2$, $b = 3$, and $c = 4$

(e) $-5xy + z$ if $x = 6$, $y = 7$, and $z = 1$

(a) $$x + 2y = (2) + 2(5)$$
$$= 2 + 10$$
$$= 12$$

(b) $$a + bc - 3 = (4) + (5)(6) - 3$$
$$= 4 + 30 - 3$$
$$= 34 \quad 3$$
$$= 31$$

(c)

$$m^2 + 4n + 1 = (3)^2 + 4(2) + 1$$
$$= 9 + 8 + 1$$
$$= 17 + 1$$
$$= 18$$

(d)

$$\frac{b + c}{7} + \frac{c}{a} = \frac{(3) + (4)}{7} + \frac{(4)}{(2)}$$
$$= \frac{7}{7} + 2$$
$$= 1 + 2$$
$$= 3$$

(e)

$$-5xy + z = -5(6)(7) + 1$$
$$= -5(42) + 1$$
$$= -210 + 1$$
$$= -209$$

Solving Simple Equations

When **solving a simple equation,** think of the equation as a balance, with the equals sign (=) being the fulcrum, or center. Thus, if you do something to one side of the equation, you *must* do the same thing to the other side. Doing the *same thing to both sides* of the equation (say, adding 3 to each side) keeps the equation balanced.

Solving an equation is the process of getting what you're looking for, or solving for, on one side of the equals sign and everything else on the other side. You're really sorting information. If you're solving for x, you must get x on one side by itself.

Addition and subtraction equations. Some equations involve only addition and/or subtraction.

Example 3: Solve for x.

$$x + 8 = 12$$

To solve the equation $x + 8 = 12$, you must get x by itself on one side. Therefore, subtract 8 from both sides.

$$
\begin{array}{r}
x + 8 = 12 \\
\underline{-8 \quad -8} \\
x \quad\;\; = 4
\end{array}
$$

To check your answer, simply plug your answer into the equation.

$$x + 8 = 12$$
$$(4) + 8 \stackrel{?}{=} 12$$
$$12 = 12 \quad \checkmark$$

Example 4: Solve for y.

$$y - 9 = 25$$

To solve this equation, you must get y by itself on one side. Therefore, add 9 to both sides.

$$
\begin{array}{r}
y - 9 = 25 \\
\underline{+9 \quad +9} \\
y \quad\;\; = 34
\end{array}
$$

To check, simply replace y with 34.

$$y - 9 = 25$$
$$(34) - 9 \stackrel{?}{=} 25$$
$$25 = 25 \quad \checkmark$$

Example 5: Solve for x.

$$x + 15 = 6$$

To solve, subtract 15 from both sides.

$$\begin{array}{r} x + 15 = 6 \\ -15 \quad -15 \\ \hline x \quad\;\; = -9 \end{array}$$

Notice that in each case above opposite operations are used; that is, if the equation has addition, you subtract from each side.

Multiplication and division equations. Some equations involve only multiplication or division. This is typically when the variable is already on one side of the equation, but there is either more than one of the variable, such as $2x$, or a fraction of the variable, such as

$$\frac{x}{3} \quad \text{or} \quad (\tfrac{1}{2})x$$

In the same manner as when you add or subtract, you may multiply or divide both sides of an equation by the same number, *as long as it is not zero,* and the equation will not change.

Example 6: Solve for x.

$$3x = 9$$

Divide each side of the equation by 3.

$$3x = 9$$

$$\frac{3x}{3} = \frac{9}{3}$$

$$\frac{\cancel{3}x}{\cancel{3}} = \frac{9}{3}$$

$$x = 3$$

To check,

$$3x = 9$$

$$3(3) \overset{?}{=} 9$$

$$9 = 9 \quad \checkmark$$

Example 7: Solve for y.

$$\frac{y}{5} = 7$$

To solve, multiply each side by 5.

$$\frac{y}{5} = 7$$

$$(5)\left(\frac{y}{5}\right) = (7)(5)$$

$$\left(\frac{\cancel{5}}{1}\right)\left(\frac{y}{\cancel{5}}\right) = 35$$

$$y = 35$$

To check,

$$\frac{y}{5} = 7$$

$$\frac{35}{5} = 7$$

$$7 = 7 \quad \checkmark$$

Example 8: Solve for x.

$$\tfrac{3}{4}x = 18$$

To solve, multiply each side by $\tfrac{4}{3}$.

$$\tfrac{3}{4}x = 18$$

$$(\tfrac{4}{3})(\tfrac{3}{4}x) = (18)(\tfrac{4}{3})$$

$$\left(\frac{\cancel{4}}{\cancel{3}}\right)\left(\frac{\cancel{3}}{\cancel{4}}x\right) = \left(\frac{\overset{6}{\cancel{18}}}{1}\right)\left(\frac{4}{\underset{1}{\cancel{3}}}\right)$$

$$x = 24$$

Or, without canceling,

$$\tfrac{3}{4}x = 18$$

$$(\tfrac{4}{3})(\tfrac{3}{4}x) = (18)(\tfrac{4}{3})$$

$$\tfrac{12}{12}x = (\tfrac{18}{1})(\tfrac{4}{3})$$

Notice that on the left you would normally not write $\tfrac{12}{12}$ because it would always cancel to $1x$, or x.

$$x = \tfrac{72}{3}$$

$$= 24$$

Combinations of operations. Sometimes, you may have to use more than one step to solve the equation. In most cases, do the addition or subtraction step first, and then, after you've sorted the variables to one side and the numbers to the other, multiply or divide to get only one of the variables (that is, a variable with no number, or 1, in front of it: x not $2x$).

Example 9: Solve for x.

$$2x + 4 = 10$$

Subtract 4 from both sides to get $2x$ by itself on one side.

$$\begin{array}{r} 2x + 4 = 10 \\ \underline{-4 \quad -4} \\ 2x \quad = \quad 6 \end{array}$$

Then divide both sides by 2 to get x.

$$2x = 6$$

$$\frac{2x}{2} = \frac{6}{2}$$

$$\frac{\cancel{2}x}{\cancel{2}} = \frac{6}{2}$$

$$x = 3$$

To check, substitute your answer into the original equation.

$$2x + 4 = 10$$
$$2(3) + 4 \overset{?}{=} 10$$
$$6 + 4 \overset{?}{=} 10$$
$$10 = 10 \quad \checkmark$$

Example 10: Solve for x.

$$5x - 11 = 29$$

Add 11 to both sides.

$$\begin{array}{r} 5x - 11 = 29 \\ \underline{+11 \quad +11} \\ 5x \quad = \quad 40 \end{array}$$

Divide each side by 5.

$$5x = 40$$

$$\frac{5x}{5} = \frac{40}{5}$$

$$\frac{\cancel{5}x}{\cancel{5}} = \frac{40}{5}$$

$$x = 8$$

To check,

$$5x - 11 = 29$$

$$5(8) - 11 \stackrel{?}{=} 29$$

$$40 - 11 \stackrel{?}{=} 29$$

$$29 = 29 \quad \checkmark$$

Example 11: Solve for x.

$$\tfrac{2}{3}x + 6 = 12$$

Subtract 6 from each side.

$$
\begin{array}{rcr}
\tfrac{2}{3}x + 6 & = & 12 \\
-6 & & -6 \\
\hline
\tfrac{2}{3}x & = & 6
\end{array}
$$

Multiply each side by $\frac{3}{2}$.

$$\frac{2}{3}x = 6$$

$$\left(\frac{3}{2}\right)\left(\frac{2}{3}x\right) = 6\left(\frac{3}{2}\right)$$

$$\left(\frac{\cancel{3}}{\cancel{2}}\right)\left(\frac{\cancel{2}}{\cancel{3}}x\right) = \left(\frac{\cancel{6}^{3}}{1}\right)\left(\frac{3}{\cancel{2}_{1}}\right)$$

$$x = 9$$

To check,

$$\frac{2}{3}x + 6 = 12$$

$$\frac{2}{3}(9) + 6 \stackrel{?}{=} 12$$

$$\frac{2}{\cancel{3}_{1}}\left(\frac{\cancel{9}^{3}}{1}\right) + 6 \stackrel{?}{=} 12$$

$$6 + 6 \stackrel{?}{=} 12$$

$$12 = 12 \quad \checkmark$$

Example 12: Solve for y.

$$\frac{2}{5}y - 8 = -18$$

Add 8 to both sides

$$\begin{array}{r} \frac{2}{5}y - 8 = -18 \\ +8 \quad +8 \\ \hline \frac{2}{5}y \quad = -10 \end{array}$$

Multiply each side by $\frac{5}{2}$.

$$\tfrac{2}{5}y = -10$$

$$(\tfrac{5}{2})(\tfrac{2}{5}y) = (-10)(\tfrac{5}{2})$$

$$\left(\frac{\cancel{5}}{\cancel{2}}\right)\left(\frac{\cancel{2}}{\cancel{5}}y\right) = \left(-\frac{\cancel{10}^{5}}{1}\right)\left(\frac{5}{\cancel{2}_{1}}\right)$$

$$x = -25$$

To check,

$$\tfrac{2}{5}y - 8 = -18$$

$$\tfrac{2}{5}(-25) - 8 \stackrel{?}{=} -18$$

$$\frac{2}{\cancel{5}_{1}}\left(-\frac{\cancel{25}^{5}}{1}\right) - 8 \stackrel{?}{=} -18$$

$$-10 - 8 \stackrel{?}{=} -18$$

$$-18 = -18 \quad \checkmark$$

Example 13: Solve for x.

$$3x + 2 = x + 4$$

Subtract 2 from both sides (which is the same as adding -2).

$$
\begin{array}{rcl}
3x + 2 &=& x + 4 \\
-2 & & -2 \\
\hline
3x &=& x + 2
\end{array}
$$

Subtract x from both sides

$$\begin{array}{r} 3x = x + 2 \\ \underline{-x -x} \\ 2x = 2 \end{array}$$

Note that $3x - x$ is the same as $3x - 1x$.

Divide both sides by 2.

$$2x = 2$$

$$\frac{2x}{2} = \frac{2}{2}$$

$$\frac{\cancel{2}x}{\cancel{2}} = \frac{2}{2}$$

$$x = 1$$

To check,

$$3x + 2 = x + 4$$

$$3(1) + 2 \stackrel{?}{=} (1) + 4$$

$$3 + 2 \stackrel{?}{=} 1 + 4$$

$$5 = 5 \quad \checkmark$$

Example 14: Solve for y.

$$5y + 3 = 2y + 9$$

Subtract 3 from both sides.

$$\begin{array}{r} 5y + 3 = 2y + 9 \\ \underline{-3 -3} \\ 5y = 2y + 6 \end{array}$$

Subtract $2y$ from both sides

$$\begin{array}{rr} 5y = & 2y + 6 \\ -2y & -2y \\ \hline 3y = & 6 \end{array}$$

Divide both sides by 3.

$$3y = 6$$

$$\frac{3y}{3} = \frac{6}{3}$$

$$\frac{\cancel{3}y}{\cancel{3}} = \frac{6}{3}$$

$$y = 2$$

To check,

$$5y + 3 = 2y + 9$$

$$5(2) + 3 \stackrel{?}{=} 2(2) + 9$$

$$10 + 3 \stackrel{?}{=} 4 + 9$$

$$13 = 13 \quad \checkmark$$

Sometimes, you will need to simplify each side (combine like terms) before actually starting the sorting process.

Example 15: Solve for x.

$$3x + 4 + 2 = 12 + 3$$

First, simplify each side

$$3x + 4 + 2 = 12 + 3$$

$$3x + 6 = 15$$

Subtract 6 from both sides.

$$3x + 6 = 15$$
$$\underline{ - 6 \quad -6}$$
$$3x = 9$$

Divide both sides by 3.

$$\frac{3x}{3} = \frac{9}{3}$$

$$\frac{\cancel{3}x}{\cancel{3}} = \frac{9}{3}$$

$$x = 3$$

To check,

$$3x + 4 + 2 = 12 + 3$$
$$3(3) + 4 + 2 \stackrel{?}{=} 12 + 3$$
$$9 + 4 + 2 \stackrel{?}{=} 12 + 3$$
$$13 + 2 \stackrel{?}{=} 15$$
$$15 = 15 \quad \checkmark$$

Example 16: Solve for *x*.

$$4x + 2x + 4 = 5x + 3 + 11$$

Simplify each side.

$$6x + 4 = 5x + 14$$

Subtract 4 from both sides.

$$6x + 4 = 5x + 14$$
$$\underline{ - 4 \qquad -4}$$
$$6x = 5x + 10$$

Subtract $5x$ from both sides.

$$
\begin{array}{rr}
6x = & 5x + 10 \\
-5x & -5x \\
\hline
x = & 10
\end{array}
$$

To check,

$$4x + 2x + 4 = 5x + 3 + 11$$
$$4(10) + 2(10) + 4 \stackrel{?}{=} 5(10) + 3 + 11$$
$$40 + 20 + 4 \stackrel{?}{=} 50 + 3 + 11$$
$$60 + 4 \stackrel{?}{=} 53 + 11$$
$$64 = 64 \quad \checkmark$$

Mathematical word problems often bring needless fear and anxiety to math students. Don't let the descriptive words surrounding the important numbers and information scare you. The following **solving process** will often help you simplify what appears to be a difficult word problem.

Solving Process

A **solving process** is a step-by-stop method to assist you in approaching word problems in an organized, focused, and systematic manner.

- **Step 1.** Find and underline or circle *what the question is asking.* Identify *what you are trying to find.* How tall is the girl? What is the cost? How fast is the car? Underlining or circling what you are looking for will help you make sure that you are answering the question.

- **Step 2.** *Focus on and pull out important information* in the problem. Watch for key words that help give you a relationship between the values given.

- **Step 3.** *Set up the work that is needed,* that is, the operations necessary to answer the question. This may be setting up a basic operation such as multiplication, setting up a ratio or proportion, or setting up an equation.

- **Step 4.** *Do the necessary work or computation carefully.* One of the most common and annoying mistakes is to set up the problem correctly and then make a simple computational error.

- **Step 5.** *Put your answer into a sentence to make sure that you answered the question being asked.* Another common error is the failure to answer what was being asked.

- **Step 6.** *Check to make sure that your answer is reasonable.* A simple computational error, such as accidentally adding a zero, can give you a ridiculous answer. Estimating an answer can often save you from this type of mistake.

Key Words

The following key words will help you understand the relationships between the pieces of information given and will give you clues as to how the problem should be solved.

- **Add**

    Addition: as in *The team needed the addition of three new players* . . .
    Sum: as in *The sum of 5, 6, and 8* . . .
    Total: as in *The total of the last two games* . . .
    Plus: as in *Three chairs plus five chairs* . . .
    Increase: as in *Her pay was increased by $30* . . .

- **Subtract**

    Difference: as in *What is the difference between 8 and 5* . . .
    Fewer: as in *There were ten fewer girls than boys* . . .
    Remainder: as in *What is the remainder when* . . . or *How many are left when* . . .
    Less: as in *A number is six less than another number* . . .
    Reduced: as in *His allowance was reduced by $5* . . .
    Decreased: as in *What number decreased by 7 is 5* . . .
    Minus: as in *Seven minus some number is* . . .

- **Multiply**

  Product: as in *The product of 3 and 6 is* . . .
  Of: as in *One-half of the people in the room* . . .
  Times: as in *Six times as many men as women* . . .
  At: as in *The cost of five yards of material at $9 a yard is* . . .
  Total: As in *If you spend $20 per week on gas, what is the total for a two-week period* . . .
  Twice: as in *Twice the value of some number* . . . (multiplying by 2)

- **Divide**

  Quotient: as in *The final quotient is* . . .
  Divided by: as in *Some number divided by 5 is* . . .
  Divided into: as in *The coins were divided into groups of* . . .
  Ratio: as in *What is the ratio of* . . .
  Half: as in *Half of the cards were* . . . (dividing by 2)

As you practice working word problems, you will discover more key words and phrases that will give you insight into the solving process.

Example 1: Jack bowled four games for a total score of 500. What was his average score for a game?

Step 1: *Find and underline or circle what the question is asking.*

What was <u>his average score for a game</u>?

Step 2: *Focus on and pull out important information.*

four games for a total score of 500

Step 3: *Set up the work that is needed.*

$$500 \div 4$$

(The total divided by the number of games gives the average.)

Step 4: *Do the necessary work or computation carefully.*

$$\begin{array}{r} 125 \\ 4 \overline{)500} \end{array}$$

Step 5: *Put your answer into a sentence to make sure that you answered the question being asked.*

Jack's average score for a game is 125.

Step 6: *Check to make sure that your answer is reasonable.*

Since four games of 125 total 500, your answer is reasonable and correct.

Example 2: Judy scored 85, 90, and 95 on her last three algebra tests. What was her average score for these tests?

Step 1: *Find and underline or circle what the question is asking.*

What was <u>her average score for these tests</u>?

Step 2: *Focus on and pull out important information.*

85, 90, and 95 on three tests

Step 3: *Set up the work that is needed.*

$$(85 + 90 + 95) \div 3 =$$

(The total divided by the number of scores gives the average.)

Step 4: *Do the necessary work or computation carefully.*

$$\frac{(85 + 90 + 95)}{3} = \frac{270}{3}$$

$$= 90$$

Step 5: *Put your answer into a sentence to make sure that you answered the question being asked.*

Judy's average test score for these tests was 90.

Step 6: *Check to make sure that your answer is reasonable.*

Since her scores were 85, 90, and 95, the average should be halfway between 85 and 95. So 90 is a reasonable answer.

Example 3: Frances goes to the market and buys two boxes of cereal at $4 each, three bottles of milk at $2 each, and two cans of soup at $1 each. How much change will Frances get from a $20 bill?

Step 1: *Find and underline or circle what the question is asking.*

How much change will Frances get from a $20 bill?

Step 2: *Focus on and pull out important information.*

two boxes at $4 each

three bottles at $2 each

two cans at $1 each

$20 bill used

Step 3: *Set up the work that is needed.*

$$2 \times 4 =$$
$$3 \times 2 =$$
$$2 \times 1 =$$
$$\$20 - ? =$$

Step 4: *Do the necessary work or computation carefully.*

$$2 \times 4 = 8$$

$$3 \times 2 = 6$$

$$2 \times 1 = 2$$

$$8 + 6 + 2 = 16$$

$$20 - 16 = 4$$

Step 5: *Put your answer into a sentence to make sure that you answered the question being asked.*

Frances will get $4 change.

Step 6: *Check to make sure that your answer is reasonable.*

Since the total expenses were $16,
then $4 change from a $20 bill is reasonable.

Example 4: Sarah can purchase a television for $275 cash or for $100 as a down payment and ten monthly payments of $30 each. How much money can Sarah save by paying cash for the television?

Step 1: *Find and underline or circle what the question is asking.*

<u>How much money can Sarah save</u> by paying cash for the television?

Step 2: *Focus on and pull out important information.*

cash $275

$100 down plus ten payments of $30

Step 3: *Set up the work that is needed.*

$$100 + (10 \times 30) =$$

$$? - 275 =$$

Step 4: *Do the necessary work or computation carefully.*

$$100 + (10 \times 30) = 100 + 300$$

$$= 400$$

$$400 - 275 = 125$$

Step 5: *Put your answer into a sentence to make sure that you answered the question being asked.*

Sarah can save $125 by paying cash.

Step 6: *Check to make sure that your answer is reasonable.*

Ten payments of $30 each is $300 plus a $100 down payment gives $400. This is $125 more than $275, so the answer is reasonable.

Example 5: If apples sell for $3.25 per dozen, how many apples can Maria buy for $13.00?

Step 1: *Find and underline or circle what the question is asking.*

how many apples can Maria buy for $13.00?

Step 2: *Focus on and pull out important information.*

$3.25 per dozen

$13.00

Step 3: *Set up the work that is needed.*

$$13.00 \div 3.25 = ? \text{ dozen}$$

$$= ? \text{ apples}$$

Step 4: *Do the necessary work or computation carefully.*

$$13.00 \div 3.25 = 4 \text{ dozen}$$
$$= 4 \times 12$$
$$= 48$$

Step 5: *Put your answer into a sentence to make sure that you answered the question being asked.*

Maria can buy 48 apples for $13.00.

Step 6: *Check to make sure that your answer is reasonable.*

Since Maria could buy 12 apples for about $3, then it is reasonable that she could buy 48 apples for about $12.

Example 6: Sequoia Junior High School has a student enrollment of 2000. If 30% of the students are seventh graders, how many seventh graders are enrolled at the school?

Step 1: *Find and underline or circle what the question is asking.*

how many seventh graders are enrolled at the school?

Step 2: *Focus on and pull out important information.*

2000 students

30% are seventh graders

Step 3: *Set up the work that is needed.*

30% of 2000 =

Step 4: *Do the necessary work or computation carefully.*

$$30\% \text{ of } 2000 = .30 \times 2000$$

$$= 600$$

Step 5: *Put your answer into a sentence to make sure that you answered the question being asked.*

There are 600 seventh graders enrolled at Sequoia Junior High.

Step 6: *Check to make sure that your answer is reasonable.*

Since 30% of 1000 is 300, then 30% of 2000 is 600.
The answer is reasonable.

Example 7: Jim Chamberlain, the center for the Westhills Basketball Stars, makes 75% of his free throws. If Jim shoots eighty free throws in a season, how many of his free throws does he make?

Step 1: *Find and underline or circle what the question is asking.*

how many of his free throws does he make?

Step 2: *Focus on and pull out important information.*

makes 75%

shoots eighty free throws

Step 3: Set up the work that is needed.

$$75\% \text{ of } 80 =$$

Step 4: *Do the necessary work or computation carefully.*

$$75\% \text{ of } 80 = \tfrac{3}{4} \times 80$$

$$= 60$$

Step 5: *Put your answer into a sentence to make sure that you answered the question being asked.*

Jim makes sixty free throws.

Step 6. *Check to make sure that your answer is reasonable.*

Since 50% of his eighty free throws would be forty, then it's reasonable for 75% to be sixty.

Example 8: Each week, John spends $50 of his income on entertainment. If John earns $200 a week, what percent of his income is spent on entertainment?

Step 1: *Find and underline or circle what the question is asking.*

what <u>percent</u> of his income is <u>spent on entertainment</u>?

Step 2: *Focus on and pull out important information.*

$50 on entertainment

$200 income

Step 3: *Set up the work that is needed.*

$$\frac{50}{200} = \frac{\text{entertainment}}{\text{income}}$$

Step 4: *Do the necessary work or computation carefully.*

$$\frac{50}{200} = \frac{1}{4}$$
$$= 25\%$$

Step 5: *Put your answer into a sentence to make sure that you answered the question being asked.*

John spends 25% of his income on entertainment.

Step 6: *Check to make sure that your answer is reasonable.*

Since 50 is half, or 50%, of 100, then 50 would reasonably be 25% of 200.

Example 9: The Gomez family spends 30% of their income for food. If the family spent $6000 for food last year, what was the family income for last year?

Step 1: *Find and underline what the question is asking.*

what was the family income for last year?

Step 2: *Focus on and pull out important information.*

30% of income for food

6000 spend for food last year

Step 3: *Set up the work that is needed.*

30% of income = 6000

$30\%x = 6000$

Step 4: *Do the necessary work or computation carefully.*

$$30\%x = 6000$$

$$.30x = 6000$$

$$\frac{.30x}{.30} = \frac{6000}{.30}$$

$$\frac{\cancel{.30}x}{\cancel{.30}} = \frac{6000}{.30}$$

$$x = 20,000$$

Step 5: *Put your answer into a sentence to make sure that you answered the question being asked.*

The Gomez family income for last year was $20,000.

Step 6: *Check to make sure that your answer is reasonable.*

Since 30% of the family income is spent for food, and 30% of $20,000 is $6,000, then the answer is reasonable.

Example 10: A miniature piano keyboard is $16\frac{1}{2}$ inches wide. If each key is $1\frac{1}{2}$ inch wide, how may keys are there?

Step 1: *Find and underline or circle what the question is asking.*

<u>how many keys</u> are there?

Step 2: *Focus on and pull out important information.*

$16\frac{1}{2}$ inches wide

$1\frac{1}{2}$ inch keys

Step 3: *Set up the work that is needed.*

$$16\frac{1}{2} \div 1\frac{1}{2} = \quad \text{or} \quad 16.5 \div 1.5 =$$

Step 4: *Do the necessary work or computation carefully.*

$$16\frac{1}{2} \div 1\frac{1}{2} = \frac{33}{2} \div \frac{3}{2}$$

$$= \frac{33}{2} \times \frac{2}{3}$$

$$= \frac{\overset{11}{\cancel{33}}}{\cancel{2}} \times \frac{\cancel{2}}{\cancel{3}}_{1}$$

$$= 11$$

Or

$$\frac{16.5}{1.5} = \frac{165}{15}$$

$$= 11$$

Step 5: *Put your answer into a sentence to make sure that you answered the question being asked.*

There are eleven keys on the miniature keyboard.

Step 6: *Check to make sure that your answer is reasonable.*

Since 16 divided by 1 is 16, and 16 divided by 2 is 8, the answer of 11 is reasonable.

Example 11: The low temperature on Big Bear mountain was 30 degrees on Monday, 20 degrees on Tuesday, −10 degrees on Wednesday, and 15 degrees on Thursday. If you total the changes in low temperature from each day to the next, what is the total number of degrees change?

Step 1: *Find and underline or circle what the question is asking.*

what is the total number of degrees change?

Step 2: *Focus on and pull out important information.*

30 on Monday

20 on Tuesday

−10 on Wednesday

15 on Thursday

Step 3: *Set up the work that is needed.*

$$30 \text{ to } 20 =$$
$$20 \text{ to } -10 =$$
$$-10 \text{ to } 15 =$$

Step 4: *Do the necessary work or computation carefully.*

$$30 \text{ to } 20 = 10$$
$$20 \text{ to } -10 = 30$$
$$-10 \text{ to } 15 = 25$$
$$\text{total change} = 10 + 30 + 25 = 65$$

Step 5: *Put your answer into a sentence to make sure that you answered the question being asked.*

The low temperature changed 65 degrees during the days Monday through Thursday.

Step 6: *Check to make sure that your answer is reasonable.*

Since the low temperature dropped 10, dropped 30, and rose 25, the total of 65 is correct and reasonable.

Example 12: The Silvarado Flash, a solar-powered land vehicle, travels at a maximum speed of 97 miles per hour. At this rate, how far will the Silverado Flash travel in 15 hours?

An important formula to remember is $d = rt$, or distance equals rate times time.

Step 1: *Find and underline or circle what the question is asking.*

<u>how far</u> will the Silverado Flash <u>travel in 15 hours</u>?

Step 2: *Focus on and pull out important information.*

> 97 miles per hour
>
> 15 hours

Step 3: *Set up the work that is needed.*

> distance = rate × time
>
> $d = 97 \times 15$

Step 4: *Do the necessary work or computation carefully.*

> $d = 97 \times 15$
>
> $= 1455$

Step 5: *Put your answer into a sentence to make sure that you answered the question being asked.*

The Silverado Flash will travel 1455 miles in 15 hours.

Step 6: *Check to make sure that your answer is reasonable.*

At 100 miles per hour for 15 hours, the Silverado Flash would have traveled 1500 miles. So the answer of 1455 miles is reasonable.

Example 13: Asaf can run around the track, 440 yards, in 65 seconds. At this same rate, how far could Asaf run in 195 seconds?

Step 1: *Find and underline or circle what the question is asking?*

> how far could Asaf run in 195 seconds?

Step 2: *Focus on and pull out important information.*

> 440 yards in 65 seconds
>
> 195 seconds

Step 3: *Set up the work that is needed.*

You could set up the proportion

$$\frac{440 \text{ yards}}{65 \text{ seconds}} = \frac{x \text{ yards}}{195 \text{ seconds}}$$

or simply divide 195 by 65 and multiply by 440.

$$195 \div 65 = ? \text{ times } 440$$

Step 4: *Do the necessary work or computation carefully.*

$$\frac{440}{65} = \frac{x}{195}$$

Since $\quad\quad 3 \times 65 = 195$

$$3 \times 440 = 1320$$

Cross multiplying would also work but would be more time consuming.

Step 5: *Put your answer into a sentence to make sure that you answered the question being asked.*

Asaf could run 1320 yards in 195 seconds.

Step 6: *Check to make sure that your answer is reasonable.*

If Asaf ran 400 yards every 60 seconds, or one minute, and since 195 seconds is just over three mintues, then three times 400, or 1200, would be a good approximation. So 1320 yards is a reasonable answer.

Example 14: A researcher tagged 100 frogs in a nearby pond. One week later she took a sample and only 5 out of 20 frogs were tagged. Using this method, how many frogs would she approximate are in the pond?

Step 1: *Find and underline or circle what the question is asking.*

how many frogs would she approximate are in the pond?

Step 2: *Focus on and pull out important information.*

100 tagged

5 out of 20 tagged

Step 3: *Set up the work that is needed.*

You could set up the proportion

$$\frac{5 \text{ tagged}}{20 \text{ total}} = \frac{100 \text{ tagged}}{x}$$

Step 4: *Do the necessary work or computation carefully.*

$$\frac{5}{20} = \frac{100}{x}$$

$$5x = 2000$$

$$\frac{5x}{5} = \frac{2000}{5}$$

$$\frac{5x}{5} = \frac{2000}{5}$$

$$x = 400$$

Step 5: *Put your answer into a sentence to make sure that you answered the question being asked.*

She would approximate that there are 400 frogs in the pond.

Step 6: *Check to make sure that your answer is reasonable.*

Since she tagged 100, and 5 out of 20, or 1 out of 4, came out tagged, then it is reasonable to have a total of 400 frogs in the pond.

Example 15: Jolia has $300 in the bank, She works at a bakery and makes $40 per day. If she deposits all of her earnings in the bank and does not make any withdrawals, how many days of work will it take for her to have $740 in the bank?

Step 1: *Find and underline or circle what the question is asking.*

<u>how many days of work</u> will it take for her <u>to have $740</u> in the bank?

Step 2: *Focus on and pull out important information.*

$300 in the bank

$40 per day

$740 total

Step 3: *Set up the work that is needed.*

Let d stand for the number of days. Then $40d$ is the amount of money earned in d days, and $40d + 300$ is the amount of money she would have in the bank at the end of d days. So

$$40d + 300 = 740$$

Step 4: *Do the necessary work or computation carefully.*

$$
\begin{array}{rcl}
40d + 300 & = & 740 \\
- 300 & & -300 \\
\hline
40d & = & 440
\end{array}
$$

$$\frac{40d}{40} = \frac{440}{40}$$

$$\frac{\cancel{40}d}{\cancel{40}} = \frac{440}{40}$$

$$d = 11$$

Step 5: *Put your answer into a sentence to make sure that you answered the question being asked.*

Jolia would have to work eleven days to have $740 in the bank.

Step 6: *Check to make sure that your answer is reasonable.*

At $40 per day, if she worked eleven days, she would have $440. Add this to the $300 she already had, and the total of $740 is correct and reasonable

Example 16: A train is 50 miles from Seattle. It is traveling away from Seattle at a speed of 60 miles per hour. In how many hours will the train be 290 miles from Seattle?

Step 1: *Find and underline or circle what the question is asking.*

In how many hours will the train be 290 miles from Seattle?

Step 2: *Focus on and pull out important information.*

50 miles from Seattle

60 miles per hour

290 miles

Step 3: *Set up the work that is needed.*

Let h stand for the number of hours. Then $60h$ is the distance traveled in h hours, and $60h + 50$ is the distance from Seattle in h hours. So

$$60h + 50 = 290$$

Step 4: *Do the necessary work or computation carefully.*

$$60h + 50 = 290$$
$$\underline{-50 \quad -50}$$
$$60h = 240$$

$$\frac{60h}{60} = \frac{240}{60}$$

$$\frac{\cancel{60}h}{\cancel{60}} = \frac{240}{60}$$

$$h = 4$$

Step 5: *Put your answer into a sentence to make sure that you answered the question being asked.*

It will take the train four hours to be 290 miles from Seattle.

Step 6: *Check to make sure that your answer is reasonable.*

At 60 miles per hour, in four hours the train will travel 240 miles. Since it is starting 50 miles from Los Angeles, the total would be 290 miles.

Example 17: A number n is increased by 25, and the outcome is 57. What is the value of the number n?

Step 1: *Find and underline or circle what the question is asking.*

What is the <u>value</u> of the number <u>n</u>?

Step 2: *Focus on and pull out important information.*

n is increased by 25

outcome is 57

Step 3: *Set up the work that is needed.*

$$n + 25 = 57$$

Step 4: *Do the necessary work or computation carefully.*

$$
\begin{array}{rcr}
n + 25 = & & 57 \\
- 25 & & -25 \\
\hline
n \quad = & & 32
\end{array}
$$

Step 5: *Put your answer into a sentence to make sure that you answered the question being asked.*

The value of the number is 32.

Step 6: *Check to make sure that your answer is reasonable.*

Since 32 plus 25 equals 57, the answer checks and is reasonable.

Example 18: The length of a Lambo Speed Wagon is 75 centimeters less than the length of a Corvette. The Lambo is 410 centimeters long. How long is a Corvette?

Step 1: *Find and underline what the question is asking*

How long is a Corvette?

Step 2: *Focus on and pull out important information.*

Lambo is 75 centimeters less

Lambo is 410 centimeters

Step 3: *Set up the work that is needed.*

Let c represent the length of a Corvette. Then $c - 75$ is the length of a Lambo. So

$$c - 75 = 410$$

Step 4: *Do the necessary work or computation carefully.*

$$c - 75 = 410$$
$$\underline{\; + 75 \quad +75}$$
$$c \qquad = 485$$

Step 5: *Put your answer into a sentence to make sure that you answered the question being asked.*

The length of a Corvette is 485 centimeters.

Step 6: *Check to make sure that your answer is reasonable.*

Since $485 - 75$ is 410, the answer checks and is reasonable.

Study Smart with Cliffs StudyWare®

Cliffs StudyWare is interactive software that helps you make the most of your study time. The programs are easy to use and designed to let you work at your own pace.

Test Preparation Guides—
Prepare for major qualifying exams.
• Pinpoint strengths and weaknesses through individualized study plan. • Learn more through complete answer explanations. • Hone your skills with full-length practice tests. • Score higher by utilizing proven test-taking strategies.

Course Reviews—Designed for introductory college level courses.
• Supplement class lectures and textbook reading. • Review for midterms and finals.

Qty.	Title		Price	Total	Qty.	Title		Price	Total
	Algebra I	☐IBM ☐Mac	19.98			Statistics	☐IBM ☐Mac	19.98	
	Algebra I CD-ROM (IBM & Mac)		24.98			Trigonometry (TMV)	☐IBM ☐Mac	19.98	
	Biology	☐IBM ☐Mac	19.98			ACT	☐IBM ☐Mac	19.98	
	Biology CD-ROM (IBM & Mac)		24.98			ACT CD-ROM (IBM & Mac)		24.98	
	Calculus	☐IBM ☐Mac	19.98			CBEST	☐IBM ☐Mac	19.98	
	Calculus CD-ROM (IBM & Mac)		24.98			College Bound Bndl. (ACT, SAT, U.S. News)		29.98	
	Chemistry	☐IBM ☐Mac	19.98			GED	☐IBM ☐Mac	19.98	
	Chemistry CD-ROM (IBM & Mac)		24.98			GMAT	☐IBM ☐Mac	19.98	
	Economics	☐IBM ☐Mac	19.98			GRE	☐IBM ☐Mac	19.98	
	Geometry	☐IBM ☐Mac	19.98			GRE CD-ROM (IBM & Mac)		24.98	
	Geometry CD-ROM (IBM & Mac)		24.98			LSAT	☐IBM ☐Mac	19.98	
	Math Bundle (Alg., Calc., Geom., Trig.)		39.98			SAT I	☐IBM ☐Mac	19.98	
	Physics	☐IBM ☐Mac	19.98			SAT I CD-ROM (IBM & Mac)		24.98	

Prices subject to change without notice.

Available at your booksellers, or send this form with your check or money order to Cliffs Notes, Inc., P.O. Box 80728, Lincoln, NE 68501
http://www.cliffs.com

get the Cliffs Edge!

☐ Money order ☐ Check payable to Cliffs Notes, Inc.

☐ Visa ☐ Mastercard Signature _____

Card no. _____ Exp. date _____

Name _____

Address _____

City _____ State_____ Zip_____

GRE is a registered trademark of ETS. SAT is a registered trademark of CEEB.